D0239514

FASTER
THAN
SOUND

Lockheed F-22A Raptor. *Lockheed-Martin*

FASTER
THAN
SOUND

THE STORY OF SUPERSONIC FLIGHT

Second Edition

BILL GUNSTON OBE

Haynes Publishing

First published in 1992 by Patrick Stephens Limited,
an imprint of Haynes Publishing.
This 2nd revised edition published in 2008 by Haynes Publishing.

© Bill Gunston 1992, 2008

All rights reserved. No part of this publication may be reproduced
or stored in a retrieval system or transmitted, in any form or
by any means, electronic, mechanical, photocopying,
recording or otherwise, without prior permission in writing from
Haynes Publishing.

A catalogue record for this book is available from the
British Library

ISBN 978 1 84425 564 1

Library of Congress control no. 2008926364

Published by Haynes Publishing, Sparkford,
Yeovil, Somerset BA22 7JJ, UK
Tel: 01963 442030 Fax: 01963 440001
Int. tel: +44 1963 442030 Int. fax: +44 1963 440001
E-mail: sales@haynes.co.uk
Website: www.haynes.co.uk

Haynes North America Inc.
861 Lawrence Drive, Newbury Park,
California 91320, USA

Dassault Mirage 2000. *PRM Aviation*

While every effort is taken to ensure the
accuracy of the information given in this
book, no liability can be accepted by the author or publishers for any loss,
damage or injury caused by errors in, or omissions from the information given.

Printed and bound in Great Britain by J. H. Haynes & Co. Ltd, Sparkford

CONTENTS

Introduction to the First Edition   ix

Introduction to the Second Edition   xi

Glossary   xiii

1.   The speed of sound   1

2.   The sound barrier   10

3.   Supersonic flight   20

4.   Through the barrier   35

5.   The first supersonic fighters   51

6.   Faster and faster   80

7.   Propulsion   96

8.   Supersonic bombers   107

9.   SSTs   124

10. V/STOL and VG   153

11. Today's supersonic aircraft   172

12. Aerospace-planes   196

     Index   230

Lockheed F-35A Lightning II. *Lockheed-Martin*

ACKNOWLEDGEMENTS FOR FIRST EDITION

For the loan of photographs I am grateful to Phillip Clark, Nigel Eastaway, Ken Gatland, John Golley, Mike Hooks, Ken Munson and Malcolm Passingham – and to Philip Jarrett, who in addition did a dedicated job as the book's editor. I am indebted to British Aerospace for granting permission to use diagrams originally prepared by British Aircraft Corporation. Finally, I thank Jenny Faithfull for painstaking picture research.

ACKNOWLEDGEMENTS FOR SECOND EDITION

Not least, I thank Jonathan Falconer and his team at Haynes for turning the printed words into an enlarged and updated book. Once again I would like to thank Philip Jarrett and this time Peter R. March for their contributions to the illustrations.

BAC Lightning F6. *PRM Aviation*

INTRODUCTION
TO THE FIRST EDITION

Man first flew faster than sound in 1947. Since then we have had tens of thousands of supersonic aircraft and, as I write, the world total of active supersonic aircraft is not fewer than 20,000. What almost got barricades put up in the streets was the suggestion, around 1970, that to this huge total should be added about one-hundredth as many supersonic transports (SSTs).

SSTs occupy just one of the twelve chapters in this book. I hope readers will find there an objective account of how man tried – and succeeded – to continue to do what he has done for thousands of years, namely to shrink the planet and make long journeys quicker and less tiresome. When you had one-sided arguments taken to imbecilic lengths it was a natural response to take the contrary view. Ceaseless bombardment with shrill cries to the effect that the SST would destroy historic buildings and kill millions tended to make one feel that such arguments were nonsensical, and lose sight of the fact that they actually contained a grain or two of truth.

This book is not specifically for or against the SST. It is not that kind of book, though doubtless the 'antis' will search its pages to find fresh ammunition now that the prospect of a next-generation SST is looming on the horizon. Personally, I would consider it almost unbelievable if, when the Concordes finally run out of hours after 40 uneventful years, nothing existed to replace them. After all, every year we replace about 1,000 supersonic military aircraft.

No, this book is just another in my long series written for people who are interested in aircraft and the technology that lies behind them. The growing number of people who are concerned about our environment often draw attention to pollution of the atmosphere (about one-millionth part of this pollution being caused by aircraft). They do not often comment on how lucky we are to have such a wonderful atmosphere, which not only sustains life but also provides oxygen for heat engines and a fluid medium in which many kinds of aircraft can fly. Without it we would literally be sunk!

It so happens that this medium transmits disturbances in the form of pressure waves we call sound. In the first chapter I discuss these waves, and also the fundamentally different waves, called shockwaves, which are caused by the passage through the air of a supersonic body. We are all familiar with the biggest shockwaves produced on our Earth, which we call thunder. Lightning strokes, which travel many times up and down faster than the eye can see, are very like localised super-fast SSTs.

In 1933 a German aerodynamicist was studying the flow over accurately made steel model wings placed downstream of a small supersonic nozzle. He reported his findings two years later at a big international congress. Nobody appeared to take the slightest interest. In Britain, for example, the notion of sweptback wings appeared to burst on us as a totally new idea discovered in material captured in Germany in 1945! As explained in this book, we were – if not at the level of working aerodynamicists, at least at the level of the top decision-takers – so utterly ignorant that we thought every supersonic aeroplane had to have sweptback wings, so we cancelled the only one we had. Everything possible was done to prevent the magnitude of this disaster from becoming apparent to the public.

One man who was particularly counselled not to rock the boat was Sir Frank Whittle. He was deeply saddened at this lunatic decision, and not only because his team was developing the engine. Way back in October 1929, as a young flying officer, he had been learning to instruct on Avro 504s, which were fabric-covered biplanes which flew at 85mph. Previously he had flown Siskin fighters with 111 Squadron, the last word in speed at up to 156 m.p.h. This gives some idea of the greatness of Whittle's mental leap in not only inventing the jet engine but also in recognising immediately that this would enable aircraft to fly not just much faster than before, but faster than the speed of sound. So one can imagine how distressed he was when, in January 1946, the British government abruptly cancelled the Miles M.52, the brilliantly conceived project which could have given the nation a lead in supersonic flight (except that I don't believe a word of it; we lacked the national will to do any such thing, and the company's enthusiasm was ignored).

So it was left to the team of Bell Aircraft, Reaction Motors, the National Advisory Committee for Aeronautics (NACA) and the USAAF to provide a rather brutish vehicle called the XS-1 for 'Chuck' Yeager to strap in to and fly faster than sound in October 1947. At the time, and for ten years afterwards, the media loved to talk about the 'sound barrier'. Many people sincerely believed that aeroplanes made a bang when they 'pierced' the sound barrier. Some thought they made a second bang when they 'unpierced' it. In the early 1950s, when sonic bangs were thought to be a novel phenomenon, I was a staff man on *Flight* magazine, and many and weird were the sonic-boom explanations we ran in our correspondence columns.

Today all this seems naive and ignorant, which is what we were. We now know that any vehicle, of any shape, can go supersonic if it is strong enough and its thrust exceeds its drag to beyond Mach 1. We know that sweepback is totally irrelevant. The XS-1 did not need it. The fastest aeroplane ever built, the X-15, has none, and most of today's supersonic fighters, such as the F-15, F-16, F-18, YF-22, YF-23, MiG-25, MiG-29 and MiG-31, merely have taper. Such aircraft are untroubled by any adverse effect such as buffet or pitch-up when they accelerate from, say, Mach 0.9 to 1.3, or let speed bleed off again. The YF-22 and -23 are among the first examples of a new species of so-called 'supercruise' fighters which, like SSTs, can cruise faster than sound for as long as the fuel lasts, and it will last much longer than before because they can do it, again like an SST, without using an afterburner to boost thrust. On the other hand, people who fly faster than sound do so in essentially straight lines. You can dogfight in close combat or you can fly at supersonic speed, but you can't do both at the same time. Air combat tends to require something around 400kt or less.

Bombers try to avoid air combat, and so in theory can go as fast as their engines will permit. Today, when the aircraft designer sits in front of a computer graphics display and pretty well dials in any cruise Mach number he wishes, it is nostalgic to look back to the early 1950s. Then, despite the fact that the technology of high-speed aircraft was galloping ahead faster than at any time before or since, the problem of sustained supersonic flight was so difficult that the designer had to cheat. It was quite possible to build a supersonic bomber provided that the target was within about 50 miles. As the customers all wanted to bomb places about 3,000 miles further away the problems were enormous, as explained in Chapter 8. In solving them, the harassed design teams laid the groundwork for future SSTs, which have to behave as perfectly normal commercial transports in all respects except speed and cruise height.

Of course, once you have gone through that non-existent barrier the sky is the limit. You may have a brush with something called the 'thermal thicket', and even a head-on confrontation with the dreadful-sounding 'heat barrier', but as time goes by we will put them all behind us. Back in February 1986 President Reagan said that the X-30A, the US National Aerospace Plane, would lead to 'a plane that will shrink travel times between Tokyo and Washington DC to less than two hours'.

One is tempted to make weak jokes. Do the Australians, for example, really want 100,000 lager-loaded Brits to descend on them just for a soccer match, and if nowhere is more than two hours away how are businessmen going to catch up on all that work they do today in 747s? Seriously, isn't the prospect of a much smaller world worth striving for? Man has tried to travel faster and more effortlessly throughout recorded history, and no amount of clever electronic communication will ever arrest the trend. We can count ourselves fortunate that, certainly in the lifetime of younger readers, 'two hours to anywhere on Earth' will be broadly an accomplished fact.

This book is a mixture of pictures of aeroplanes and aerospacecraft, and diagrams and graphs which give an insight into supersonic flight. I hope the two sit happily together; you can at a pinch ignore the hard stuff. Incidentally, all gallons are Imperial. Regarding the photos, I have tried to avoid too many hackneyed ones.

There's a lot to write about in tracing the story of supersonic flight; it's obvious that each chapter could be a book in itself.

Bill Gunston
Haslemere
February 1992

INTRODUCTION
TO THE SECOND EDITION

The original edition appeared at the end of a wonderful era of flight, one unlikely to be repeated. It was wonderful because from 1902 until 1992 it had been natural to build new aircraft that were faster than their predecessors. Today it is more difficult to find the thrill of thrusting ever further up the scale of Mach numbers. Halfway through the 90 years carefully selected test pilots were for the first time penetrating what was popularly regarded as a 'sound barrier'. Within a very few years all but the poorest air forces had squadrons of supersonic fighters, and the richest counries even had supersonic bombers. By the 1960s several countries were working on SSTs (supersonic transports), and in 1976 the Anglo-French Concorde entered regular scheduled airline service. At a stroke this effectively shrank our planet to half its previous size. Crossing the North Atlantic took three hours instead of seven.

What nobody then could have predicted is that for various reasons, most having little to do with the technology, Concorde did not become the first of the many. When I wrote the original book it was generally taken for granted that, when the Concordes were retired, a bigger and longer-ranged successor would take over. Astonishingly, in October 2004 the Concordes were flown off to museums, and were not followed by any successor. The considerable number of people who had routinely crossed the North Atlantic twice in a day found that this was no longer possible. For the first time in the history of life on Earth, the speed of the fastest passenger vehicle had suddenly been reduced by more than 60 per cent, with no sign of anything being done about it.

What is perhaps equally surprising is that for 50 years the fastest fighters have got slower and slower. Nobody today would dream of making a successor to the 2,446mph Republic XF-103, and several of the newest fighters cannot exceed Mach 1.6 or 1.8. This is partly explained by the fact that flying at speeds higher than this means that you travel in an almost straight line. Another explanation is that the fighter has become a mere carrier of weapons, and the outcome of the battle is decided by radio waves or infra-red.

It does not seem natural for humans to give up a capability, once it has been achieved. The X-30A, mentioned in the Introduction to the First Edition, was never built. It has been followed by a succession of later NASA (US National Aeronautics and Space Administration) vehicles, so far extending to X-50, all in some way intended to further mans ability to navigate in and beyond our atmosphere. Many similar projects exist in Europe, Japan and a few other countries. Half a century ago the emphasis was on travelling ever faster. This is still an important objective, and young people today may well live to reach Australia from England, or vice-versa, in two hours. The difference is this is now just one of several important objectives, and a poll of our planet's population would probably find that going long distances in less time comes a long way down the list.

Bill Gunston, OBE, FRAeS
Haslemere
March 2009

Lockheed-Martin F-22A Raptor. *PRM Aviation*

GLOSSARY

Absolute temperature   Temperature measured relative to absolute zero; the latter is close to -273°C, so normal room temperature (15°C) is 288°K absolute.

ACT   Active controls technology, in which a vehicle's flight-control system automatically and continuously issues commands to counter atmospheric turbulence, or any other undesired input.

Afterburner   Also called a reheat jetpipe, in which fuel is burned downstream of a turbojet or turbofan engine in order to increase the jet's temperature and velocity.

Air-breathing   Relying on oxygen from the atmosphere for combustion.

AOA   Angle of attack, the angle at which a wing meets the oncoming air.

Area rule   A fundamental law for minimum transonic drag, stating that cross-section areas plotted from nose to tail should rise smoothly to a maximum and then diminish smoothly back to zero (any wings, external stores, engines, etc. being included).

Aspect ratio   A measure of the slenderness of a wing, numerically equal to span squared divided by area.

Booster   An engine or vehicle used purely to accelerate a payload to high speed, e.g. to orbital velocity.

Boost/glide   A vehicle with no internal propulsion, boosted to high (e.g., orbital) velocity, and thereafter gliding back through the atmosphere.

Boundary layer   The layer of fluid (e.g., air) in contact with a moving body; the thickness of this layer grows from zero at the nose to perhaps several inches (say, 80mm) 200ft or more (say, 60m) further aft.

CCV   Control-configured vehicle, an aircraft whose design is dictated by the acceptance of inherent instability (e.g., a wish to fly tail-first) in order to achieve higher manoeuvrability and other advantages.

CG   Centre of gravity, through which the resultant of a body's weight forces acts.

Chord   Distance across a wing or tail surface from leading edge to trailing edge.

Con/di   A jet-engine's propulsive nozzle, initially convergent, to the throat, and then divergent in profile, used only in supersonic flight.

CP   Centre of pressure, the line or point through which the resultant of all wing lift forces acts.

Critical Mach number   The Mach number at which shockwaves first appear at the point (or, in a real body, the line) of peak suction; also defined as the Mach number at which such shockwaves first exert a noticeable influence on an aircraft.

Delta   Shaped more or less like a triangle.

FBW   Fly by wire, a flight-control system in which the pilot's commands are transmitted in the form of varying electrical impulses.

Hypersonic   Several times faster than the local speed of sound; normally taken to mean Mach 5 or above.

ICBM   Inter-continental ballistic missile, defined in 1954 as having a range of at least 5,500 nautical miles (6,325 miles).

IRBM   Intermediate-range ballistic missile, defined in 1954 as having a range of at least 1,500 nautical miles (1,727 miles).

L/D   The ratio of the total lift to the total drag of a vehicle.

LEX   A chordwise extension to the leading edge; LERX adds 'root'.

Mach cone   The envelope of all the shockwaves emanating from a supersonic aircraft at the cone's vertex.

Mach number   Ratio, expressed as a decimal, of a vehicle's speed to that of sound in the surrounding medium.

Mach wave   A shockwave of exceedingly small amplitude caused by supersonic flow past a point, or a sharp or thin leading edge.

Normal shock   A shockwave perpendicular to the direction of flow, peculiar to Mach 1.

Re-entry   Flight from space back through a planetary (especially the Earth's) atmosphere.

Reheat   British term for afterburning.

Root   The junction of a wing or tail surface with the body.

Semi-span   The distance from the aircraft's centreline to either wingtip.

Slender delta   A delta wing of very low aspect ratio, in which the root chord is much greater than the span.

Slew wing   A wing made in one unit from tip to tip, but pivoted to the rest of the aircraft at the centre.

Soaking   Flying at a high supersonic speed for long enough for the entire aircraft to settle down at a high temperature (except for specially cooled parts).

TAS   True air speed.

T/c   The ratio of maximum thickness of a wing to its chord at the same spanwise station.

VC engine   Variable-cycle engines incorporate a large valve enabling part or all of the airflow to be shut off to convert the engine to or from a turbofan/turbojet to a turboramjet, or even a pure ramjet.

VG   Variable-geometry, able to change shape in flight.

Vortex lift   Extra lift generated by powerful vortices spiralling back from a sharp leading edge across a slender delta wing.

Bristol 188. *PRM Aviation*

Chapter 1

THE SPEED OF SOUND

Towards the end of the Second World War, in the mid-1940s, when fighter pilots dived steeply to very high speeds they often encountered unexpected, and often frightening, control difficulties. The effects included a sudden buffet, resembling a violent shaking of the aircraft, and either a tendency to pitch up, threatening to pull the wings off or, more often, to dive ever more steeply no matter what the pilot might do. Many pilots made no such reports, because they had dived straight into the ground.

It all seemed mysterious and frightening. The cause was obviously the changed distribution of flows and pressures around an aircraft as it neared the speed of sound. One of the pioneer researchers into high-speed airflow was an Austrian, Ernst Mach (this rhymes with the composer Bach) who lived 1838–1916. Ever since, his name has been used as a measure of high-speed aerodynamics, the speed of sound being called Mach 1. Maddeningly, although aerodynamicists could test models in wind tunnels up to about Mach 0.8 (0.8 of the local speed of sound), and also at Mach numbers higher than about 1.2, there was no way tests could be done over the tricky and unknown transonic region from Mach 0.8 to 1.2 or thereabouts. This was because the tunnel was found to 'choke'. The presence of the model locally speeded up the flow as the air rushed to get past, so that shockwaves would begin to block the tunnel at around Mach 0.8. Thus, over 60 years ago, the only way anyone could explore transonic flight was by diving real aeroplanes. And a lot of pilots, with a lifetime of subsonic experience, were killed trying to push through this unknown frontier.

In a standard work (*High-Speed Aerodynamics*, by Dr W.F. Hilton, Longman, 1952) it was suggested that, to achieve higher speeds, aeroplanes would retract their engine, tail and even fuselage, leaving an all-wing machine which might well have variable sweepback. Beyond doubt the aerodynamicists of the immediate postwar era would be amazed to go aboard an A380, with a fuselage 27ft high and over 23ft wide, a wing root over 12ft deep, an external tail, and four external engine pods each 12ft in diameter, and then ride with 600 others at 625mph They would be equally surprised to see the apparently normal-looking wing of a 1,500mph F-15, or the shape of a 2,200mph Lockheed SR-71, or a 17,600mph Shuttle orbiter. But we must not get cocky; the future will present us with plenty of fresh unknowns.

So in this first chapter I will set the scene by discussing the gas in which we fly, and the problems with which it confronted us as we sought to move through it significantly faster than before.

We are fortunate that our planet has this atmosphere. It sustains life, protects us from harmful radiation, and does many other things, such as enabling us to speak to each other, and to fly. Air is a mixture of gases, by far the most important being nitrogen (in dry air at sea level about 78 per cent) and oxygen (about 21 per cent). The mixture is viscous, or 'sticky', in that work has to be done to make its molecules slide past each other. This is important in how aircraft fly, and how wings lift, but can largely be ignored in this book. It is also compressible, and this we cannot ignore. If air were incompressible, any small disturbance would travel through it at an infinite speed. In other words, if a body were to move through such a fluid, the resulting disturbances would be felt instantaneously throughout the whole mass of fluid. With such a fluid, the flow around any moving body would vary according to the body's shape, but would be the same for every speed.

A real fluid, however, is compressible. Small disturbances, or pressure waves, travel through real fluids at a speed which, for any particular specimen of fluid, is

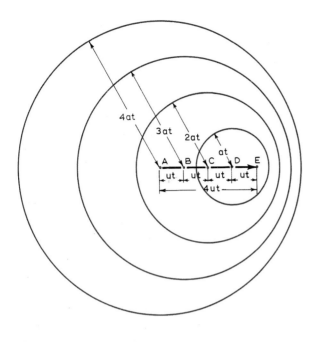

Pressure waves radiating from a small particle travelling at subsonic speed. Each ring is really a sphere.

fixed. We call this speed the speed of sound. In very loose language, the more compressible the fluid, the slower does sound travel through it. In a simplistic way, we can picture the fluid as consisting of an infinite number of tennis balls, all with space between them and free to move in any direction, or past each other. It does not matter that, if the fluid is air, four balls out of every five are marked 'nitrogen', and almost all the rest 'oxygen'. If we exploded a small grenade somewhere in the mass of fluid all the nearest balls would be blown radially outwards. When they hit other balls they would rebound, while the balls they hit would carry on transmitting the disturbance. Thus, the disturbance would expand outwards in all directions, the wave front being spherical.

How far the wave would travel would depend on the energy released in our original bang. If the original disturbance was a human whisper, the energy would be extremely small, the pressure difference at the source of the wave being typically 10^9 atmosphere, and the disturbance would be hard to detect 3m (10ft) from its source. With ordinary speech the initial dP (pressure difference) might be 10^6 atmosphere (one-millionth of an atmosphere), so the sound might be heard 20m (66ft) from the source. In every case the wave takes the form of a very small rise in pressure followed by an equal and opposite fall in pressure, which in the analogy of our tennis balls corresponds to their initial bunching up and impact followed by rebounding and widely separating,

resuming their normal average separations after the disturbance has died down.

The rate at which the balls (or gas molecules) move out and hit each other depends on what they are made of. The speed of sound is not the same for all materials; sound travels several times faster in water than in air, and faster still through metals, for example. The speed also depends slightly on the frequency of the sound, and to a very high degree on the temperature of the medium. The hotter the material, the more our tennis balls are vibrating wildly before the sound is heard, and the more rapidly does the wave-front travel.

I am anxious to avoid deterring readers with technicalities, but they may like to know that, for gaseous fluids, a (the speed of sound) is given by the formula $a^2 = \gamma RT$. In other words, the speed of sound is numerically equal to the square root of γ (the Greek letter gamma, the symbol for the ratio of the specific heats for the particular gas) multiplied by R (the gas constant) multiplied by T (the absolute temperature). If we get all our units right, then, for dry air, γ is near enough to 1.4, R is $3,087\mathrm{ft}^2$ sec^2 °C, and T at room temperature is 288°. This gives us that a (the speed of sound in such air) is the square root of 1,244,678, or 1,115.6ft/sec (340m/sec, 760.6mph). Suppose we climb into the stratosphere, where the temperature is -56.5°C, or in absolute terms 216.5°. In such intensely cold air $a^2 = 935,670$, so that a becomes only 967.3ft/sec (294.8m/sec, 659.4mph). Thus, there is a big difference between the speed of sound at sea level and at high altitude, or between the Sahara and a Siberian winter. When jet aircraft were up against the so-called 'sound barrier' (Chapter 2), those trying to break speed records sought the hottest air they could find.

Any real aeroplane usually flies through air which is pretty much the same in all directions (though, of course, it will normally get warmer towards the Earth and, at below FL361 (36,100ft), colder towards higher altitudes). We can take it that the air in the immediate vicinity of our aircraft is the same everywhere. Accordingly, the disturbances created by the aircraft's passage through the air keep travelling away from it in the form of spherical wave-fronts. Again we have to simplify things, and instead of a real aeroplane we imagine the disturbance as coming from a point in space. Thus, from this point there ceaselessly radiates a succession of spherical waves, which in cross-section (if the point were to hover motionless) resemble the ripples on a pond into which a pebble has been dropped. In three dimensions they look like the layers of an onion.

But our real aeroplane is not hovering, but moving forward. Accordingly, the successive spherical waves are not concentric; instead, each emanates from a point a short distance in front of the last one. We can represent the situation by a simple figure (bearing in mind that the wave fronts are not really 2D rings but 3D spheres). In this figure the successive positions of our (point-sized) aeroplane are $A/B/C/D/E$, each position being t seconds later than its predecessor. We call our airspeed u, and the speed of sound a (it doesn't matter what units we choose, so long as they are the same, for example ft/sec). The figure shows the situation at time E. The wave emitted at this time has not yet left position E. The wave formed at D now has a radius of at, centred on D. Similarly we have a sphere of radius $2at$ centred on C, one of $3at$ centred on B, and one of $4at$ centred on A. We can easily get an idea of the cruising Mach number, because this must equal ut/at, or $4ut/4at$, which can be seen to be about 0.7. So this is a picture of the noise from a Mach 0.7 aircraft, such as a BAe 146.

Now let us board a supersonic aircraft and open the throttles a little. Eventually we reach an airspeed where $u = \underline{a}$, which we call Mach 1. In this special case the nose of the aircraft continuously travels forward at exactly the same speed as the fronts of the waves caused by its disturbance. The picture would thus resemble an endless series of spheres (far left, drawn as circles, for simplicity), all of which pass through a single point, which is the instantaneous position of the aircraft. Clearly, this is a different situation. The front of the first wave is the first warning the atmosphere ahead gets of the oncoming aircraft, but, when this warning is received, the aircraft is already there. This means that, at Mach 1, the air cannot be forewarned and smoothly pushed out of the way, as it can when u/a (Mach number) is subsonic. Instead it gets an extremely sudden shock by finding the aircraft's nose or leading edge already pushing its way through without warning. We can also see that the wave which travels along with the nose of the aircraft must be rather special, because it contains all the infinite number of sound waves generated during the aircraft's flight. We do, in fact, call such a wave a shockwave, or shock. (A weak shock, caused by a very pointed or thin body, is called a Mach wave.) A shock can be thought of loosely as an extremely intense sound wave, but in fact it exhibits significant differences.

As the diagram below shows, a sound wave has the form of a smooth sine wave, gently reaching a peak

Whereas a normal sound wave has a smooth sine-wave shape, a Shockwave reaches its peak value almost instantaneously. It is also much more powerful. Of course, the sound could be continuous (only one cycle is shown), unlike the shock, which is a single wave.

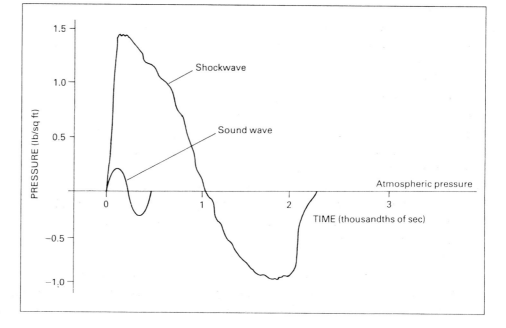

position value (the balls all colliding) and then dropping to an almost equal negative value (the balls bouncing apart, causing a rarefaction). It is these waves, alternately rarefactions or exceedingly small-amplitude compression waves, that travel at 'the speed of sound'. In contrast, the shock has a finite amplitude, always many times greater than that of a normal sound wave. Moreover – and this is crucial – instead of smoothly rising to its peak value, it attains the value almost instantaneously. In other words the wave front, on a plot of pressure against time, is almost a vertical line. As we shall see, the actual shape of this pressure/time plot, and the pressures reached, both positive and negative, depend very much on the moving body's size, shape and speed. As one might expect, we get a much bigger shock from a larger blunt body moving at 10,000mph than from a small needle-nosed one moving at 1,000mph.

So far we have not considered supersonic speeds; and this is the point at which to mention three key words that will recur often in this book. Subsonic flow is airflow where all velocities are less than the local speed of sound. It makes no difference whether the velocity is measured between a moving body and initially still air, as in the flight of an aircraft, or between moving air and a stationary body, as in the test of a model in a wind tunnel. The phenomena and the numerical measures are exactly the same. In subsonic flow there will be sound waves, sometimes quite intense ones, but no shockwaves.

The second word is transonic. This literally means 'across the speed of sound'; in other words, it is a regime in which the local Mach number is less than 1 at some points and more than 1 at others. Wherever the local Mach number exceeds 1 there are bound to be shockwaves, whose strength will depend on the size of the body around or through which the air is flowing, and the angular distance (degrees) through which the flow is turned or diverted. Loosely, the transonic regime is likely to extend from Mach 0.8 to 1.2.

The third word is supersonic. Here the local Mach number exceeds 1 everywhere; there is no subsonic region. For completeness I will throw in two more words. Hypersonic describes flow at Mach numbers several times greater than 1; there is general agreement a hypersonic aircraft flies at a Mach number of at least 5. Ultrasonic has nothing to do with speed at all, but means that a sound is pitched so high that the human ear cannot hear it.

Clearly, there must be differences between subsonic and supersonic flow, and these differences are far more than just ones of degree. Aerodynamicists have written large textbooks to explain the differences, some of which seem amazing. I don't want to put anyone off with maths, but many aerodynamic formulae contain the term (M^{2-1}),

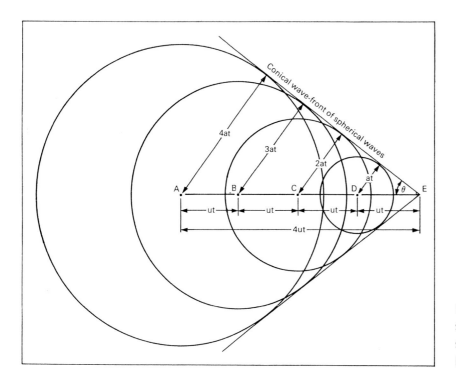

Pressure waves radiating from a small particle travelling faster than sound. It tows an expanding cone behind it.

The acute shockwaves left by a small sphere travelling through Freon gas at Mach 12.15.

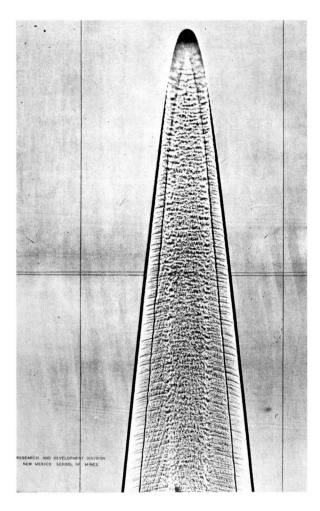

where M is Mach number. Obviously, for any given Mach number less than unity, such as 0.999, this expression will be negative, whereas for any supersonic Mach number it will be positive. This is one factor in understanding why the two types of flow are so different. I will defer this discussion until Chapter 3, which describes how air flows around and through aircraft at various speeds.

In our brief look at the propagation of sound waves from aircraft we studied what happens at M 0.70 and at M 1, but what happens at supersonic speed? The next figure shows sound waves left behind by a small particle travelling at Mach 1.65. In other words $u/a = 1.65$. Thus, t seconds after point D the pressure wave emitted at the point has formed a sphere of radius at, but our speeding particle has punched through it and gone on ahead a distance ut to reach point E. The same is true for the earlier locations A, B and C. When we add all the waves together we find they form a huge cone, with its vertex always at the moving particle, the source of the waves. We can imagine our tiny particle speeding through the sky, forever towing this expanding cone behind it.

This cone is called the Mach cone. The disturbance, the noise and turbulence caused by our speeding particle must forever lie inside this cone. Above, below or ahead of the cone, the air is completely undisturbed, and has no knowledge of the body's presence. As our body is extremely small, it causes only a very small disturbance. Such weak pressure waves always move at the speed of sound a, and if they come from a supersonic source they are, as I said earlier, called Mach waves. We can see from the diagram that the semi-angle θ of the cone depends on the ratio of a to u. In fact, the sine of the angle is numerically equal to a/u, which is $1/M$. Thus, if our particle is travelling at Mach 1.65, Sin θ must be $1/1.65$ or 0.606, and if you look this value up in trigonometry tables you will find the Mach angle is 37°18. Alternatively, if you can measure the Mach angle it is simple to work out the Mach number.

It is probably obvious that at Mach 1 the Mach angle is 90°, which is the situation we had in the previous figure, where all the spherical wavefronts stayed at the source, all moving along together at the same speed.

Likewise, at extremely high hypersonic Mach numbers, such as the small sphere travelling at Mach 12.15 shown in the photograph, the Mach angle θ is very small. To work it out, $1/12.15 = 0.0823$, and our table of sines tells us the angle must be about 4°43'. Thus we can see that the shockwave pattern left by a spacecraft re-entering the Earth's atmosphere is totally different from that of a (barely) supersonic fighter, where the Mach angles are not far from 90°.

So far I have been careful to describe our supersonic body as something extremely small, in order that it should generate only very weak Mach waves. But aircraft are quite big, and they generate not weak Mach waves but shockwaves. We have already seen that a shockwave has a form that is quite different from a sound wave. The latter always has the form of a smooth sine wave, never varying during the wave's progress outwards from its source except that its amplitude falls, eventually dying away to zero. Thus, we can identify a person's voice at any distance, almost to the point where the

A beautiful photograph of the canopy and wing shockwaves formed by a Grumman F-14A at Mach 0.9 with wings fully aft. Condensation in the expansion region makes the supersonic area visible. The air then passes through the shockwave at the rear of this region and is instantly heated; the water evaporates and ceases to be visible.

voice is too faint to be heard. A trumpet heard a mile away is still a trumpet. But pressure waves of finite (i.e. large) amplitude are totally different. They start off with a pressure difference typically thousands to millions of times greater than that of a sound or Mach wave. They travel somewhat faster than the local speed of sound (the latter being the speed of small-amplitude waves). Not least, during the initial part of their travel they change in form, almost as if the wave was being pushed impatiently from behind. Instead of rising smoothly to a peak value, as in a normal small-amplitude wave, the shockwave tries to overtake itself, and soon forms a wave with a very abrupt – virtually instantaneous – front, reaching the peak value immediately. Some workers have likened this process to the way an oncoming sea wave changes from being a normal wave, with a gradual increase in height, to being a 'breaker', with a sudden discontinuous 'step' in the height of the sea.

Thus, a shockwave is not just an intense sound wave. The latter can have many kinds of aural qualities, so that we can distinguish a note played on a clarinet from the same note played on an oboe. The shockwave is a sudden crack, bang or, if originally large but heard from a great distance, a boom. The figure opposite shows a strong shockwave being towed through the sky by an aircraft. As it has a large dP (pressure difference) it travels at velocity a', which is a little faster than a. Accordingly, the semi-angle of the shock cone θ' is slightly greater than that of θ for the same flight Mach number. Of course, the reader will also appreciate that both θ and θ' will vary slightly with atmospheric temperature. For any given value of M, the cone will be more pointed at high altitude than at sea level, and in winter than in summer.

We thus have a good picture of what happens as a body moving through a fluid, such as air, accelerates to supersonic speed. But there is much more to it than just the generation of shockwaves. The most fundamental change, of crucial importance to aircraft designers, is that when the velocity of the body relative to the air exceeds about 500mph the differences in pressure that occur round the body rise enormously, so that the compressibility of the air becomes important. At 200mph, for example, the maximum increase or decrease in pressure at any point around the body cannot exceed 5 per cent of the local atmospheric pressure. This has a negligible effect on the air density or flow pattern, and so the compressibility of the air could be ignored by aircraft designers of the 1930s. But at 400mph the pressure difference can reach 21 per cent of the atmospheric pressure, at 500mph 31 per cent, and at 600mph 51 per cent. From here on we are deep into the realm of compressibility, where the changed density of the air must be taken into account. At Mach 1.1 the pressure difference can exceed local atmospheric pressure; at Mach 2.2 it reaches about 10.7 times local atmospheric pressure, and at Mach 3 no less than 36.7 times! Obviously this exerts a colossal effect on the air's density, temperature and flow pattern.

I have already explained that, until the 1950s, it was impossible to test anything in a transonic wind tunnel.

But since the mid-nineteenth century researchers have known that it is possible to generate a supersonic jet by letting compressed air, or other gas, escape through a con/di nozzle (a nozzle which is initially convergent and then divergent). Such nozzles were used from about 1870 by the great Swedish steam-turbine engineer G.C.P. de Laval, and to this day they are called Laval nozzles. In his pioneer experiments with air, Ernst Mach let air escape from a bottle at about 100lb/sq in, and found that he could produce a jet moving at almost Mach 2. Subsequent workers let such jets blast past tiny model wings. Among the greatest of the researchers into supersonic aerodynamics were Betz, Busemann, Meyer and Prandtl, who were German, Ackeret, who was Swiss, and Glauert, Stanton and G. I. Taylor, who were British. Of course, their apparatus had to be extremely small. None of them could afford the kind of equipment we are used to today. Moreover, all research had perforce to be based on the results of previous tests, which led to mathematical formulae which were always suspect and often significantly in error. Each worker brought a little extra knowledge. They did this mostly in the period 1907–35, and readers can think for themselves how far removed was their research from the kinds of aeroplanes that actually existed at that time. Indeed, the gulf was so great that most of the pioneer supersonic literature went totally unread by any of the world's aircraft designers. It did not seem to bear the slightest relevance to their everyday problems. This was to prove very unfortunate,

and to lead to several silly decisions in the years immediately following the Second World War.

I have often wondered when the speed of sound first became of interest to humans. Obviously, the fact that sound travels quite slowly – about five seconds for a mile – must have been noticed thousands of years ago. There are countless instances one can think of where this fact must have been apparent. Anything that makes a sudden sound, and which can also be seen to happen, demonstrates that the light and sound from the source do not travel at the same speed. Sometimes one has to allow for this. In World War Two I often found myself drilling smart squads of cadet pilots. Every command has a prefix, which tells what is coming. This can be quite drawn out, and its timing is not critical. It is followed by a short sharp bark of an order. For example: Abo-o-o-u-t.... TURN! The word TURN must be given as every left foot comes to the ground. On a giant parade ground you have to know precisely when to shout the word so that it gets there as those left feet hit the ground; often it is as the preceding *right* feet hit the ground. In the same way, if you are a timekeeper at a 50m sprint swimming race you are at least 50m from the starter. Problems can be caused if some timekeepers start on the flash of the gun while others wait for the bang, because a 50m event may be decided by hundredths of a second, and the bang will take about 16 hundredths to travel the length of the pool (you pray the electronic timing will not fail!).

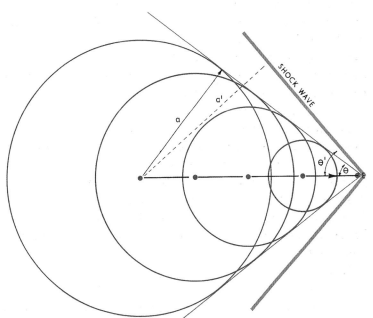

Pressure waves radiated by a large object, such as an aeroplane, travelling at supersonic speed.

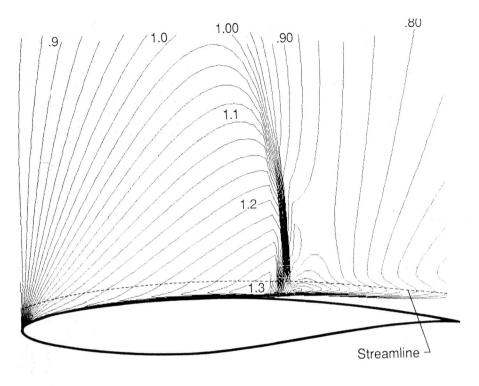

An accurate computer print-out of Mach contours for flow past an RAE 2822 aerofoil at Mach 0.75. The shockwave at the rear of the supersonic region is actually much thinner than it looks. It is causing the flow to break away from the wing.

What things were supersonic in the ancient world? The answer is, not many. Beyond doubt, gas and vapour in volcanic eruptions can issue from such pressure that it can exceed Mach 1 once it reaches the atmosphere, but this is uncommon. In any event, people in the vicinity would not have been likely to hang about taking measurements! Really, there is only one common natural source of shockwaves, and that is lightning. We think of a lightning flash as being a single arc of DC electricity from cloud to Earth, but in fact each stroke comprises several distinct flashes. Each is a law unto itself, but what may happen is that the initial stroke from cloud to Earth carves out a minimum-resistance path through the intervening atmosphere, this path thus becoming ionised. This forms a conductive path for the next stroke. This travels upwards, is by far the biggest, and carries most of the energy. Then several further strokes may follow almost the same path in each direction. It is all over in a split second, and it leaves the path not only ionised but also intensely hot. It is difficult to assign a diameter to a flash of lightning, but various erudite people have suggested 4in to 6in (100–150mm). Through this path travels electrical energy at the rate of several million horsepower. It is almost like detonating a chain of hundreds of explosive charges. The obvious result is to send out a strong shockwave, which begins

as a tube a few inches in diameter and anything up to several miles in length and spreads at velocity a' in all directions. The air collapsing back again sends out further shocks, though if you are extremely close to the source – i.e. where the lightning 'struck' – you hear only one cataclysmic bang. At a distance, thunder decays to various booms and rumbles, many of which are echoes, or arrive by different paths.

Lightning and thunder predate man on Earth by some millions of years, but I think it is only in this century that the awesome sound has been understood to be a shockwave. As it has been estimated that the lightning/thunder phenomenon is estimated to occur on our Earth not less than 100 times per second, representing the continuous expenditure of 4,000 million horsepower, one wonders why there was so much fuss about the SST.

Perhaps the earliest man-made shockwaves came from the tips of whiplashes. The dynamics of whips are complex, but with a long thong it is not difficult to 'crack' it ten times out of ten. The sharp crack is caused by the free end exceeding Mach 1. Certainly whips have been cracked for thousands of years. In contrast, arrows and other missiles were fired from cannon, muskets and other firearms were all subsonic until about the mid-nineteenth century. In engagements at sea, where distances might be considerable and visibility unimpeded except by smoke,

A US Army 155mm howitzer in action in 1944.
US National Archives

it was normal to see the flash of an enemy gun, wait several seconds and then hear the boom; a few seconds later the shot would arrive. But after 1850 some of the latest guns, smoothbore or rifled, achieved supersonic muzzle velocities (m.v.). The 68pdrs of HMS *Warrior* (1861) have m.v. of no less than 1,550ft/sec (present tense, because this battleship still exists). In the same year, the Admiralty began replacing these guns by the Armstrong 110pdrs, but as this gun had an m.v. of only 1,056ft/sec it was actually less powerful.

So far as I know, no research was done into the aerodynamics of shells or bullets until the twentieth century, and even then the only method seems to have been to take spark photographs. These are taken by a camera with an open shutter, the otherwise completely dark scene being briefly illuminated by an extremely bright but short-duration spark produced by a sudden high-voltage discharge of a condenser (capacitor). This technique was in use before 1920, and there are countless photographs of (for example) bullets slicing through playing cards or shells piercing armour. Today a typical technique is to use a xenon flash triggered by a microphone. Designers of bullets and shells were, and are, far more concerned with the projectile's stability in flight, avoiding tumbling end-over-end, and with its penetration of its targets, than with details of shockwaves, airflow or aerodynamic drag.

Thus, by the Second World War, a handful of aerodynamicists and mathematicians had laid the groundwork for a study of supersonic flow, and supersonic projectiles fired from guns had become common. But aircraft designers, as a class, knew nothing about all this, nor were they interested. Moreover, nobody had been able to study *transonic* airflow, as explained in the next chapter.

Chapter 2

THE SOUND BARRIER

One of the reasons why nobody in any aircraft company had seriously studied transonic flow before the Second World War was that there seemed to be no need. Aircraft designers tend to be pretty stolid people, with both feet on the ground, and if an Air Ministry Requirement called for a maximum speed of 'not less than 350mph and if possible 375mph' then that is precisely what they would aim at. I chatted with large numbers of designers from 1945 onwards, and almost to a man they admitted they had never contemplated such things as shockwaves or Mach numbers, until in 1945 they found captured German research material to be full of them. Even then, many of the diehards (in Britain, if not in other countries) tried to brush it all off as outlandish, cloud-cuckoo-land stuff having no bearing on reality. They were soon to learn the error of their ways.

A leader of this school of disbelievers was Sydney (later Sir Sydney) Camm, of Hawker Aircraft. Perhaps he occasionally put on an act, but in the immediate postwar period, at least until late 1947, he quite seriously professed to scorn all the newfangled Mach nonsense. It is easy to be wise with hindsight, but in fact he, more than most people, ought to have given a thought to it about ten years earlier. At that time, in 1937, he had the Hurricane in full production, and about to enter service. It was the right fighter at the right time, and is the main reason I do not serve German masters! At Kingston, Camm was beginning work on its successor, to Specification F.18/37. This gave rise to two aircraft, the R (Rolls-Royce) version becoming the Tornado and the N (Napier) version the Typhoon. Only the Typhoon went ahead, and after two years of trouble it matured as a most valuable ground-attack aircraft. It never did become a good air-combat fighter, but by sheer good fortune in 1944 aircraft that could destroy

Panzer divisions were far more valuable than good dogfighters.

I mention all this because, if anyone took it into his head to dive a Typhoon steeply, he very soon encountered buffeting sufficiently violent to make him pull out. Accompanying the buffeting was a loss of lateral control, and applying a large side force on the stick tended to roll the aircraft in the opposite direction, the aircraft exhibiting aileron reversal. These quite frightening effects were caused by the local formation of shockwaves. In other words, in places the flow around the diving Typhoon was transonic.

Nobody had even considered such a thing in 1937, and it was late 1940 before the problem really became inescapable. It was impossible to run into the trouble in level flight, because Typhoons could only nudge 400mph at between 18,000 and 21,000ft. But in a steep dive it was only a few seconds before TAS (true airspeed) reached 500 to 525mph, and here things began to happen. The root cause was that, in order to obtain sufficient spar depth for the required wing strength, and sufficient volume for the fuel, landing gears and four cannon, Camm had made the wing very thick. The crucial factor, which will occur often in this book, is thickness/chord ratio (t/c). If the chord (the distance from leading edge to trailing edge) is 20ft and the depth is 2ft then the t/c is 10 per cent, which for the past 50 years has been a typical value for propeller-driven aircraft. The t/c Camm selected for the Typhoon was no less than 18 per cent. For various reasons, he picked a profile that was not only extremely thick but also had a bluff leading edge, the point of maximum thickness occurring at 22 per cent chord (i.e., less than a quarter of the way back from the leading edge).

Now, imagine what happens when such a wing flies. To get past the wing, the air has to speed up, and it is the job

Even when carrying no bombs or rockets, the thick-winged Typhoon fighter could run into compressibility problems.

of the wing designer to try to make this acceleration occur over the top of the wing rather than on the underside. The greater the acceleration of the air, relative to the wing, the greater the reduction in pressure; it is this upper-surface suction that provides the lift that keeps the aircraft in the sky. The amount by which the flow is forced to speed up depends on the wing's t/c ratio; the thicker the wing, the greater the acceleration. And the speed-up is also greater for a wing that has its maximum thickness well forward.

In the late 1930s the National Advisory Committee for Aeronautics (NACA) in the United States began introducing so-called laminar-flow (meaning smooth and without turbulence) wings. Their t/c ratio was usually not very different from that of previous wings, but the maximum thickness occurred much further back. For

example, such a wing was fitted to the Mustang, and here the maximum thickness varied from 37.5 to 38.5 per cent of the chord, almost twice as far back as in the Typhoon.

The idea of the laminar-flow wing is that, so long as the air is accelerating across the wing, it will stay laminar. Once the maximum thickness is reached, the air slows down again. As it slows down, its pressure naturally rises, and the air flowing against an adverse pressure gradient tends to break away from the wing's surface. This causes a layer of turbulence, in contact with the wing everywhere downstream of the thickest point. The layer of air in contact with the aircraft is called the boundary layer. A turbulent boundary layer has much higher drag than a laminar one. For 60 years designers have tried (so far without very much success) to make aeroplanes that have a laminar boundary layer all over. But to return to our fighter wings, to a rough approximation the Mustang had high drag over 62 per cent of the wing, while the Typhoon had high drag over 78 per cent, and this meant that (assuming the aircraft were otherwise roughly similar) the Mustang should be faster than the Typhoon for equal power. But there was more to it than that, as we shall see.

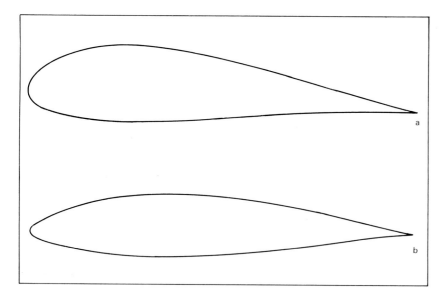

A sketch comparing the traditional wing of the Typhoon (a) with the so-called laminar profile of the Mustang (b).

So now we can imagine our Typhoon building up speed in a dive, with the air being accelerated violently over the front 22 per cent of the wing. The pilot would have no idea of the speed beyond what his ASI (airspeed indicator) told him, and, because of the low air density at about 30,000ft, this dial might tell him some figure like 300mph If he had time, he could work out mathematically that his indicated airspeed (IAS) corresponded to a true airspeed (TAS) of about 500mph, which happened to be the authorised dive limit for the Typhoon. But that would be the average, or free-stream, true airspeed. In places the air would be speeded up in order to get past, and at a TAS of 500mph the air flowing past some parts of the Typhoon would reach over 660mph, which at 30,000ft is about the speed of sound. The most important area would be the line of peak suction across the front top of the wing, from tip to tip.

When we study transonic and supersonic airflow in the next chapter we shall see that, ignoring what may happen at small bumps or constrictions, it is usually along the top of the wing, at or close to the isobar joining the points of peak suction (lowest pressure, highest velocity) that the local Mach number first reaches unity. Small shockwaves appear, and in some atmospheric conditions they can be seen with the naked eye. As speed continues to increase, they grow higher and firmer, gradually merging and forming one strong shock. At the same time this migrates aft across the wing, always marking the boundary between supersonic flow in front and subsonic flow behind. The change in conditions as the air passes through the shock is virtually instantaneous, because the air is moving very fast, and the shockwave is extremely thin. Its thickness is, in fact, roughly the same as the mean free path between the molecules of nitrogen and oxygen forming the air, and this is in the region of 0.0001in (0.0025mm). I will go into this in more detail in the next chapter.

This is rather cart before horse, because I am describing transonic airflow over a wing, whereas in the Second World War all that the test pilots had were the symptoms. These symptoms varied from one aircraft type to another, and even between individual aircraft of the same type, but they always included buffeting (high-frequency aerodynamically induced vibration, usually in the vertical plane) and either a loss of effectiveness of some or all of the flight controls, or even a reversal of effect. Frequently the symptoms included a strong pitching moment, either nose-up or nose-down. Not least of the frightening symptoms was the feeling that the dive was getting steeper and steeper, and that hauling back on the stick was having no effect. I have done this myself, but after the war, when proper recovery procedures had been worked out. It was genuinely frightening for the wartime test pilots, who didn't know whether there were any recovery procedures. One of the pilots who did most to get the Typhoon cleared for operations was R. P. 'Bee' Beamont, and he said it was every bit as exciting to dive at 450mph IAS at 25,000ft (which is about 675mph TAS, far beyond the limit authorised for Service pilots) 'where it shook with compressibility buffeting and lost elevator effectiveness' as to fly the same aircraft on low-level ops against the enemy!

With hindsight, one marvels that so thick a wing could have been designed for a powerful fighter – which Hawker Aircraft calculated in 1938 would reach 464mph in level flight – without anyone considering the problems of shockwave formation. By 1940 the need for a thinner wing was self-evident, and this eventually led to the excellent Tempest. This had a wing with slightly greater area, and t/c ratios varying from 14.5 at the root to 10 per cent at the tips. This made all the difference. Whereas in the Typhoon buffet set in at about Mach 0.64, in the Tempest there was no problem until about 0.73–0.74, a Mach number very unlikely to be reached in normal operations. Thus, merely by making the wing thinner, a fighter's unexpected transonic problems were eliminated.

Perhaps I ought to go back again to the so-called laminar-flow wing, which began to be considered by several workers at the NACA Langley Aeronautical Laboratory, in Virginia, in 1937. Langley's special low-turbulence tunnel began testing such an airfoil in June 1938. The laminar wing has a totally different profile from the traditional kind, with a less bluff front portion and the point of maximum thickness much further aft. The measured drag coefficients were about half the lowest ever recorded for a traditional wing of the same t/c ratio, and this investigation was the starting point for research into low-drag wings, which has continued to this day. In the past 50 years this research has led to what are popularly called 'supercritical' or 'rooftop' sections, specifically designed for minimum drag at high subsonic Mach numbers, as explained in Chapter 3. But the original 'laminar' sections had nothing to do with speed, beyond enabling aircraft to go faster on the same power. This was to be thrown into sharp focus in 1944, when test pilots began diving wartime fighters to the very limits, as I shall explain later.

I have happy memories of 683 hours in Harvards. This was probably the best advanced trainer of the Second World War period. Though you had to treat it with respect, it was hardly the sort of aircraft to be fitted with a Machmeter, but everyone who knows anything about Harvards knows that, as it passes you on takeoff, you hear a deafening loud rasping sound. In 1940 E. A. 'Chris' Wren, in writing the text to go with his 'Oddentification' cartoon, commented: 'I shall deliberately refrain from any reference to the shocking, ear-splitting inferno of sound which emanates from its engine and airscrew when passing by'. His cartoon (page 15) shows instructor and pupil obviously agonising in this 'inferno of sound'.

In fact, they were the lucky ones. On take-off you opened the throttle, carefully kept straight, and noted the rpm needle steady on 2,000. The Wasp gave a deep throaty rumble, and even with the canopy open there was absolutely no 'inferno of sound'. So why the problem? The answer lay in the fact that the unusual noise, heard by the people outside the aircraft, came from the tips of the propeller. Most aircraft of this family had a two-blade Hamilton Standard propeller with a diameter of 9ft. Thus, each tip travelled about 28ft 3in on each revolution. At 2,000rpm this gave a tip speed of some 943ft/sec due to rotation alone. But a propeller blade is a miniature wing, and the air flows around it in a very similar way, being speeded up as it accelerates over the cambered front lifting face. For a mean speed of 943ft/sec you can be quite certain that, even on a hot day at sea level, shockwaves will form over the last few inches of the blades, and they are the cause of the noise. There's more to it than that. Early Harvards were cleared to 2,250rpm, giving a mean tip speed of 1,060ft/sec. Moreover, even with the aircraft at rest and the blades in fine pitch, you must add something like 200 to 300mph slipstream velocity, which means that all the flow around the outer portions of the blades will be supersonic.

Thus, an aeroplane that we normally cruised at about 125kt was probably the first to experience local supersonic airflow, even if that was confined to quite small areas on the propeller. Incidentally, when I returned to Britain after years in Southern Rhodesia, I was surprised to find British Harvards not only painted yellow, and with 'British' instrument panels, but also with cropped propellers to reduce or eliminate the noise problem. In Southern Rhodesia, on a hot day at an airfield height of 4,680ft, that would have meant a pretty long take-off.

High-Mach trouble is the chief reason why piston-engined aircraft have only just exceeded 500mph, and that with great difficulty and the use of fantastic Merlin, Wasp Major, Centaurus and Double Wasp engines that must have been on the verge of blowing their cylinder heads off. At such high speeds you need extremely coarse propeller pitch, and even then the outer parts of the blades will be supersonic and generating more noise than thrust. Fairey Aviation used to display in the foyer of their Hayes factory one of the fixed-pitch Reed propellers made for the Supermarine S6 and S6B Schneider Trophy-winning seaplanes. The coarseness of the pitch was startling, and of course at take-off it tended to churn up the air rather

Right: 'Chris' Wren's *Oddentification* cartoon of the Harvard concentrated on what today we call aural signature.

Opposite: Lockheed test pilot Milo Burcham (later the first man to fly the XP-80) tells C. L 'Kelly' Johnson about the effectiveness of the P-38 dive recovery flap, seen open. A notice warns: 'Accidental operation of flap may cause loss of arm'

than give thrust. In general, propellers are fine for low speeds, but not a good idea at over 450mph Supersonic propellers are discussed in Chapter 4.

But in the Second World War, except for a handful in the final year, every power pilot flew with propellers. Wartime competition naturally made everyone, not only test pilots but also operational pilots, explore the limits. Some could do this better than others, and by late 1943 at least 20 RAF fighters, and several of the growing number of USAAF fighters in Britain, had simply dived at high speed into the ground. So far as I know, it was not until September 1942 that a proper effort began to be made to try to discover the cause.

Anyone who watched Hollywood movies in the 1930s will gain the mistaken impression that test pilots spent all their time making terminal-velocity dives, to see if the wings would come off in the stressful pull-out. But from 1942 the research into high-speed dives was not so much concerned with structural strength as with airflow, shockwaves and unknown control phenomena. At the Royal Aircraft Establishment (RAE), at Farnborough, most of this work was done with Spitfires, but it was to be 1944 before really interesting results were achieved. Meanwhile, not only the Typhoon but also most of the new crop of American fighters were experiencing problems.

Probably the first of these was the Lockheed P-38 Lightning. As early as August 1940 difficulty was being experienced by this aircraft in high-speed dives, with buffet over the tail, a tendency for the dive to steepen, and a frightening loss of elevator effectiveness. A side issue was that at very high speeds the ailerons locked almost solid, but that was the case with most fighters of the period. By 1942 it was ascertained that the turbulence was due to poor airflow around the junction between the wing and the central nacelle, and with the NACA's help

the problem was eventually cured. But the tendency for a dive to become steeper, and the difficulty of recovering to level flight, were by 1942 definitely known to be due to local flow over the inner wing becoming supersonic, the shockwaves causing turbulence and disturbed airflow over the horizontal tail (one report rather delightfully said: 'A hole is caused in the air round the tail'). Two famous test pilots then took the matter in hand. The first was Lt-Col Cass Hough, who did more than anyone else to make the hardware of the British-based 8th Air Force work. On 17 September 1942 he made a TV (terminal-velocity) dive in a P-38F Lightning, which possibly reached a higher speed than man had travelled at previously. I am not sure I believe the figures, but the dive is alleged to have begun at 43,000ft (that's hard to believe, for a start) and to have reached 525mph indicated at 35,000ft, which I simply refuse to accept. An IAS of 525mph at 35,000ft equates to a true airspeed of about 956 mph, which is nonsense. Anyway, Hough managed to pull out by winding on nose-up trim, which in the circumstances could have been lethal. As it was, the wings stayed on, and the P-38 zoomed up almost vertically before longitudinal control was restored and level flight resumed. It is a pity that we don't have meaningful figures.

In his book *L-1011 and the Lockheed Story* (Aero Publishers, 1973), which supposedly tells the true history of Lockheed, the author, Douglas J. Ingells, says of the P-38: 'It was the first plane to exceed 400mph and the first aircraft to encounter the phenomenon of "compressibility" ... As the speed was increased to 400, 500, 600mph, air piled up on the leading edge of the wing ... that's what happened one day high over the San Fernando Valley, when test pilot Ben Kelsey ran into 'compressibility' and the ship disintegrated ... His report led engineers to make the addition of a dive brake on the underside of the wing ... which whipped the problem.

Just in time, too, for the P-38 was already going into production when World War Two broke out . . .'

Of course, the P-38 was neither the first aircraft to exceed 400mph nor the first to experience compressibility, air certainly does not 'pile up', and Ben Kelsey's dives were in 1944. Incidentally, back in February 1939 he had flown the original XP-38 prototype on its famous transcontinental dash to New York, the aircraft ending up a twisted wreck on a golf course. Now, according to legend, in his 1944 dives Kelsey achieved (I quote Bill Green) 'an indicated speed of more than 750mph'. In fact (and I studied *Pilot's Notes* before once flying a P-38) pilots were explicitly warned never to go beyond a particular curve of IAS at particular heights, all corresponding to about Mach 0.73. Thus, at 35,000ft, the maximum IAS permitted was not 750mph nor about 525mph, but about 267mph. Now we are getting back into the realm of sanity. With 267mph (or 232kt) on the clock at 35,000ft, you would be well into the buffet and compressibility regime.

As for Mr Ingells' 'dive brake', that is just what it wasn't. The P-38 did not need extra drag, but something to change the flow pattern around the wing and, at the same time, to cause a nose-up pitching moment. The answer was a dive flap, or dive-recovery flap, on the underside of the outer wing, just beyond the engines. Each was a plain reinforced strip, about 5ft long and 5in chord, normally flush with the undersurface at the main spar. To recover from a dive they were opened electrically by pilot action, rotating down 35° in less than a second. This was just enough to give the required change in flow and trim, but without any significant effect on speed.

Another fighter that ran into compressibility problems was the Republic P-47 Thunderbolt. I never flew one, but this massive machine had a reputation for diving much better than it climbed. This is rather ungenerous to what became an outstanding multi-role fighter/bomber, but it is a fact that by late 1942 several P-47s, in Britain with the 56th Fighter Group as well as in the USA, had dived into the ground at high speed. Again Cass Hough was assigned to get answers, and on 27 February 1943 he dived one of the first P-47Cs to arrive in England. He started at 39,000ft and pulled out at 18,000ft. Later the US 8th Air Force put out a release saying he had 'exceeded the speed of sound'. It would have been more correct to say that the airflow over parts of the P-47C had indeed exceeded the speed of sound. Here, again, the answer was a dive-recovery flap. As on the P-38, this was a plain strip normally flush with the underside of the wing and electrically rotated to 35°. It was further back than on the P-38, being immediately ahead of the flaps.

No such flap was fitted to any of the RAF's wartime fighters. I think such an addition would have benefited the Typhoon, but most of the chaps who were killed diving Typhoons suffered structural failure, usually separation of the tail, not inability to recover to level flight. In any case, I think we were slow to learn about compressibility problems. It was not until late 1942 that the RAE tackled the problem in earnest. And perhaps the chief reason for this is that, by sheer chance, we had in the Spitfire a fighter with staggeringly good high-Mach properties. This, like the multi-spar wing of the DC-3, was purely fortuitous. When Reginald Mitchell designed that odd elliptical wing, with all its strength in

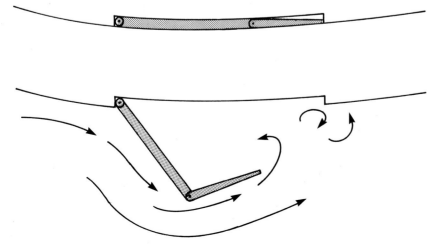

On the Lockheed P-38J-25 and subsequent production aircraft Lockheed added dive recovery flaps under the outer wings. Normally flush with the under surface, they could be quickly opened by pressing a button on the control wheel.

Tony Martindale was a test pilot of uncommon determination and courage. Twice he dived Spitfires beyond 0.9. On this occasion he brought the PR Mk XI back to Farnborough minus the complete propeller and gearbox.

the D-nose ahead of the spar, he certainly never gave a thought to shockwaves or Mach numbers.

Of course, when he did this, in 1935, he had never heard of the new low-drag laminar profiles, and picked a particular favourite of his from the traditional library of aerofoils. It was, in fact, related to that used previously on the clumsy Supermarine F.7/30, but it was dramatically thinner. Mitchell repeatedly emphasised his wish to achieve the lowest t/c ratio possible, and it was to achieve this end that he adopted the unusual structure. The t/c ratio he finally picked was 13 per cent at the root and 6 at the tip. Even the tailplane had a t/c ratio of only 9 per cent. This was a vital legacy Mitchell left when he died in June 1937. At first, nobody paid much attention to this, and it was not until 1942 that, following dives done at the RAE, it was realised that the Spitfire suffered less from compressibility problems than any other wartime aircraft, jets included.

Having said that, it is typical of the ironic paradoxes so common in aerospace to record that K5054, the original Supermarine 300 prototype, did in fact have compressibility problems. Following its first flight on 5 March 1936, it was found that the maximum level speed at the best height (about 18,000ft) was not the calculated 367mph but only 335mph, and at 17,000ft. One reason, which it took years to do much about, was that at that time engine powers were calculated using formulae derived at Farnborough which were actually shockingly optimistic. A further factor, likewise completely unsuspected previously, was that the big fixed-pitch propeller, even though it was geared down from the engine, was running into compressibility trouble at the tips. The high tip speed was in this case due more to the forward speed of the aircraft than to high rotational speed. The cure was to try out different propellers with modified tips. By 1945 Spitfires were an astonishing 100mph faster in level flight, without propeller-tip problems.

In the first series of RAE tests in 1942–3, a Spitfire IX was dived by Sqn Ldr J. R. Tobin. It achieved a true Mach number of 0.893, while neither a P-47B nor a Mustang I could get beyond 0.805. Not the least startling thing is that above Mach 0.67 the Spitfire, with its traditional wing, showed lower overall CD (drag coefficient) than the Mustang, with its new laminar profile. Equally amazing is that the very first issue of *Pilot's Notes* for the

Spitfire IX, issued in 1942, stated that the limiting Mach number (i.e. the value not to be exceeded in service) was 0.85. As no Machmeter was fitted, a graph was provided showing IAS plotted against altitude corresponding to 0.85. I'll return to this in a moment.

I would like briefly to break off here to comment that we in Britain did have several people who were aware of transonic problems and did their best, with essentially no tools, to find the answers. They included Handel Davies at Farnborough and M. J. Lighthill at the National Physical Laboratory. P. W. S. 'George' Bulman, veteran test pilot for Hawker Aircraft, made a special study of compressibility problems, and was even able to visit the USA where he could learn more about them. On returning in early 1943 he took the unprecedented step of issuing a 16-page booklet, *Piloting Techniques at Compressibility Speeds*. For the first time, it told pilots what they might expect and how they should cope. This little primer on a new subject almost certainly saved both aircraft and lives.

By 1944 it was fully recognised that the high-Mach performance of the Spitfire was exceptional. To explore this further, Sqn Ldr A. F. (Tony) Martindale made a series of dives in one of the lowest-drag versions, a PR Mk XI. This gained in having a retractable tailwheel, no guns and a smoothly curved Perspex windscreen, but its larger oil tank gave it a deep-jowled chin, which caused more violent flow acceleration and produced stronger shockwaves earlier. By 1944 plenty of Spitfires were available with two-stage Griffon engines, much faster on the level than the Mk XI, but the latter may have been able to dive faster. Martindale's first dive took place on 27 April 1944. He rolled inverted at 40,000ft and dived at full power to about 27,000ft. With the propeller fully coarse, the rpm still went 'off the clock', and the Merlin's propeller gearbox disintegrated, most of it departing along with the propeller. Unbelievably, the cowling panels remained, and Martindale made a perfect landing back at Farnborough. Subsequent analysis showed the peak Mach number to have been 0.92, or about 625mph On a later dive Martindale's engine blew up and caught fire at about Mach 0.9; undeterred by dense low cloud, he made a crash landing near Woking. Then, despite a spinal injury, he returned to the burning wreck to retrieve the recording camera. Later dives were by Flt Lt Mair.

This is absolutely staggering. Mach 0.92 is still higher than anything reached by any other piston-engined aircraft, and higher than the limit for such jet and rocket aircraft as the Me 163, Me 262, Meteor, Vampire, Venom, Sea Hawk, F-80, F-84, F-89, and F9F Panther. Of course, to reach such a figure you had to climb as high as possible, do a wing-over and dive almost vertically at full power, praying that the propeller would stay fully coarse. At around 0.9 the Spitfire was not exactly manoeuvrable; it might tend to roll left or right, and was strongly nose-down. This would have been frightening but for the fact that, as you dived into denser air, the ambient temperature increased rapidly, constantly increasing the local speed of sound and reducing the Mach number. Thus, having reached over 0.9 at perhaps 30,000ft, you would be down to 0.75 at lower levels, and able to recover to level flight. A nice point is that, by 1944, the aircraft used by the RAE for high-Mach tests were fitted with Machmeters. These were originally little more than pitot/static ASIs (airspeed indicators) with an inbuilt correction for absolute temperature, and calibrated in M (Mach number). Jeffrey Quill, the chief test pilot on the Spitfire programme throughout the war, wryly observed that the new instrument 'was extremely useful on the Mustang, but as it was calibrated only up to 0.8 it was off the scale during a good part of the Spitfire dives'.

There is yet more irony. In 1943, Supermarine's design team, under Mitchell's successor Joe Smith, were busy creating a completely new wing, which, fitted to a Spitfire (Mk VIII or XIV) would meet specification F.1/43. By 1944 this had led to a new fighter, the Spiteful. The new wing had an area of only 210sq ft, compared with the standard Spitfire's 242sq ft, and of course it had one of the new low-drag laminar profiles. This did result in lower drag over much of the speed range, but at the cost of various quite severe problems which took a long time to cure. What was totally unexpected was that the high-Mach performance of the new wing was inferior to that of the old Spitfire. After the war the Spiteful wing was used almost without change (apart from removal of the cooling radiators) for the company's first jet, the E 10/44, which led to the naval Attacker. I have no hesitation in claiming that this would have done better with the original wing of the Spitfire!

Amazingly, as late as 1942 many chief designers were still ignorant of the problems they would encounter if they fitted fast aircraft with thick wings. One of the last British fighters designed during the Second World War was the Westland Welkin. This was intended specifically for interceptions at very high altitude, where high-Mach problems would be encountered earlier. It first flew on 1 November 1942 and it was soon clear that high-speed dives were out of the question; the thick (19.15 per cent) wings, of over 70ft span, tried to vibrate themselves to pieces. Compressibility trouble was also experienced with one type of propeller. The thick wings, in my opinion, made the Welkin useless except for intercepting aircraft that could neither dive away nor manoeuvre. Westland Technical Director Teddy Petter was later to create such excellent aircraft as the Canberra and the Lightning

In many branches of science there is a time delay in communicating new knowledge to what might be termed 'workers at the coal face', such as aircraft designers. Thus, many aspects of supersonic flight had been explored before 1930, for example by the famous aerodynamicists listed in Chapter 1. One of these workers, Adolf Busemann, working with the Mach 1.5 tunnel at Göttingen, explored the pressure and force coefficients on a thin two-dimensional wing. At the Fifth Volta Conference in Rome in 1935 he presented a much more accurate second-order theory of supersonic flow around a wing He also presented various suggestions for how a supersonic aircraft might be designed. One of these, the concept of a biplane in which the shockwaves from the upper and lower wings mutually interfere with each other in a favourable manner, has never proved practical. The other idea was to sweep back the wing. He showed how, by mounting the wing at a sweep angle of 35° or 40°, the air would behave as if the wing was actually thinner, and the effective Mach number would be reduced.

Another worker at the Kaiser Wilhelm Institute for Flow Research at Göttingen was Albert Betz. In 1935 he picked up the swept-wing idea and spent the next four years exploring it in detail. He published several papers culminating in a major treatise in 1939. Presumably we in Britain not only had nobody who ever visited Göttingen (a leading world centre of aerodynamic research) but didn't even bother to read their openly published reports!

It seems quite beyond belief that this extremely important new idea of sweepback should not only have 'sunk without trace', but should not even have been remembered when, during the Second World War, the high-Mach behaviour of aircraft began to be of the most immediate practical importance. The popular explanation is that the transonic advantage of sweepback was somehow 'forgotten' or 'not then understood'. For example, Dr Hilton's 1952 textbook says, 'It was not until later that the full significance of the possibility of obtaining subsonic force coefficients and high efficiency at low supersonic speeds by sweeping back the wing was fully understood.' He must have known that Busemann and Betz had spelt it out in open literature, ten years earlier.

As we shall see, sweepback is of major interest only in the transonic regime. If you are going to fly at sustained supersonic speed then, though you will have to have the lowest possible t/c ratios for your wing and tail, there is little advantage in sweepback. Swept wings are needed to enable subsonic aircraft to cruise at higher Mach numbers. They have also been used for some aircraft with brief supersonic 'dash' capability, such as the Lightning (which had the exceptional sweep angle of 60°), but the truly supersonic aircraft almost never has a swept wing. Some have wings of extremely low aspect ratio (chord very large in relation to the span), examples being Concorde and the SR-71, but the latest, such as the F-22 and F-35, have wings whose plan shapes would not have looked unusual in the Second World War.

I commented on the gulf that separates the worker in the laboratory from the people who actually design or operate aircraft. Until the Second World War, the vast world of supersonic aerodynamics – where, as we shall see in the next chapter, things often happen in completely the opposite way to subsonic flows – was a realm totally alien and unknown to the entire world of practical aviation. High Mach numbers could be attained only in wind tunnels, and for reasons of cost these often had working sections hardly bigger than a matchbox. As we shall see, it was particularly difficult to put anything inside these tunnels, and absolutely impossible to investigate the crucial transonic regime between about Mach 0.8 and 1.2. There was no such difficulty when it came to testing real free-flying aircraft, but there was no means of propelling them at the desired speeds. The first rocket aircraft, in June 1928, is reputed to have taken one minute to cover three-quarters of a mile, which equates to a TAS of 45mph When, a year later, young Frank Whittle invented the turbojet, the aircraft he was actually flying, the Avro 504N, was capable of 100mph flat out. Nothing in the RAF squadrons could reach 200mph Small wonder that his invention generated little interest. But it was eventually to change the scene completely, and make supersonic knowledge of the most immediate practical importance.

Chapter 3

SUPERSONIC FLIGHT

We have now seen how, once aeroplanes began to reach high subsonic Mach numbers, beyond about 0.70, they ran into frightening new phenomena. Aircraft whose handling was immaculate in level flight suddenly turned into malevolent demons in steep dives. If more people had read some of the hundreds of learned papers then available on transonic airflow around wings and bodies there would have been less of a problem. As it was, pilots and even designers felt they were up against unknown and awesome forces which, for no apparent reason, could assail aircraft with what felt like hammer blows, shake them violently, and make ailerons or even elevators work in the opposite way to that intended. No wonder that everyone involved began to talk of a 'sound barrier'. It certainly seemed as though the idea of building an aeroplane that could 'pierce' this notional barrier was a pipe-dream.

I will outline how this was done in the next chapter. First, we need one more chapter of theory before we get to the aeroplanes. Of course, the reader can skip it, but it is written in simple language.

In Chapter 1, I commented on the fact that the mathematical expression M^2-1 occurs in many of the equations of aerodynamics. So, too, does $1-M^2$. Both these terms pass through zero at $M=1$ (the local speed of sound) and then change sign, in the first case from negative to positive and in the second from positive to negative. Among other things, this completely alters the way an airflow behaves. To anyone unfamiliar with aerodynamics, even subsonic flow may seem add. For example, consider the venturi. This is the name given to a tube which contracts to a minimum cross-section, called the throat, and then expands again back to its original size. If you asked a non-aerodynamicist how the pressure would vary as the air flowed through it he or she would probably say: 'It starts off at atmospheric pressure, builds up to maximum pressure in the throat and then, once past the constriction, falls back to atmospheric pressure at the exit'. This seems common sense. If we replaced the air molecules by people, then the crowd, shoving and pushing to get through, would certainly suffer highest pressure at the throat. What actually happens with an airflow is the exact opposite.

One of the first things a student of aerodynamics learns is Bernoulli's theorem. This is a statement of the conservation of energy in a fluid flow. Broadly, and making various assumptions (such as ignoring the effects of gravity or friction), it states that any fluid in steady motion must always have the same total energy per unit mass. Thus, if no energy (such as heat) is added or subtracted, then the sum of the static and dynamic pressures must be constant. In other words, if speed (dynamic pressure) increases, then the static pressure must decrease. As air flows through a venturi, the speed clearly rises to a maximum at the throat and then falls back to the original level. Thus, to keep total energy constant, pressure must *fall to a minimum at the throat* and then rise back to atmospheric pressure. Venturis are often used to provide suction to drive gyro instruments. If one imagines a venturi cut through and opened out into a straight line, it is possible to see how a wing lifts. As in the venturi, the air is accelerated across the top of the wing, the reduction in pressure (in effect sucking the wing upwards) being proportional to the amount by which the flow is speeded up.

Having rather laboriously explained what happens when air flows through a venturi, this must now be qualified by saying that everything is totally different if the flow is supersonic. Here the air entering the venturi does not accelerate but slows down, so its pressure increases, reaching a maximum at the throat. Once through the

Simple sketches showing the variation in pressure (manometer U-tubes) and velocity (dials) as air flows through a venturi. The results may seem to contradict common sense. In subsonic flow the pressure is lowest at the constriction, while the supersonic flow accelerates through an expanding nozzle

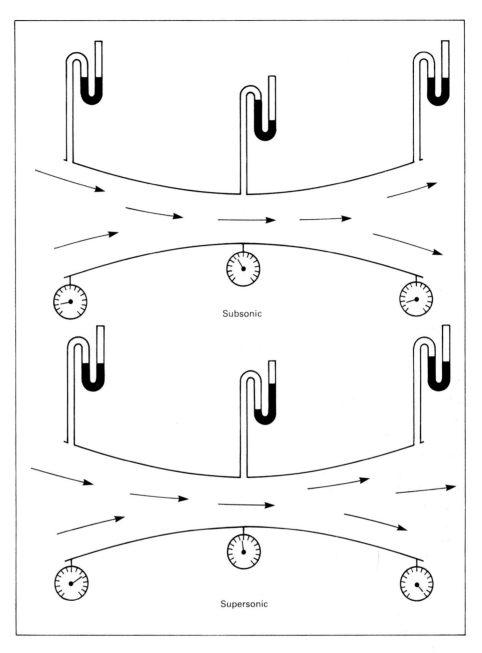

Subsonic

Supersonic

throat, the air accelerates and rarefies – in other words its density falls. Once we have taken on board the startling idea that subsonic and supersonic flows are so totally different, there is just one further possible cause of confusion. I have avoided the word 'expansion'. We often think of expansion as the opposite of contraction, but we must remember that if air flows at subsonic speed along an expanding duct then it actually compresses – its pressure increases. At supersonic speed there is no problem, because the flow expands and rarefies – it falls in both density and pressure.

We can now look at supersonic flow around noses, bodies and wings. We have already seen how a theoretical point moving at supersonic speed causes a Mach cone, with its vertex at the source. Downstream, the weak Mach waves gradually dissipate away to nothing, just as do ordinary sound waves. We have also seen how a practical body, having finite dimensions, causes a shock cone, the semi-angle of which is slightly greater than that of the Mach cone from the point source. It does not need deep thought to see that, if we could move a fine wire edge-on at supersonic speed, we should g

6/2198654

a cone but a wedge formed by two planes intersecting at the wire. If, instead of a wire, we had a real wing, then we should get two strong shock planes, again forming a wedge and intersecting at or near the leading edge.

If the wing is thin, the Mach number well above unity, and the leading edge sharp, then these shockwaves will be formed exactly at the leading edge. We say they are 'attached' to the wing. If the wing is thicker, the leading edge rounded, and the Mach number not very much above unity, then the shockwave stands off ahead of the leading edge. We call this a 'detached' shock. In the early days of research into how to design supersonic aircraft it was almost always assumed that their wings would have to have really sharp leading edges. Such an airfoil is extremely poor at take-off and landing, and almost all of today's supersonic aircraft have more normal rounded leading edges. Supersonic missiles can be accelerated violently, and so can have thin sharp-edged wings.

Having looked at what happens when a supersonic airflow passes through a contraction or an expansion, we can investigate how such a flow travels over a wing, or any other body. If it travels past a straight wall, parallel to the streamlines, as along most of the untapering length of an SR-71 or Concorde, then nothing much happens to it apart from the usual boundary-layer effect close to the solid surface. As in subsonic flow, a supersonic boundary-layer flow is slowed down relative to the surface of the aircraft, and there is little we can do about it. In a B-52H or a Harrier, for example, the scrubbing of supersonic propulsive jets against a solid surface causes a significant

fall in propulsive efficiency, but that is incidental. We are interested in supersonic flows whose direction is changed by the aircraft. The result is either a compressive flow or an expansive one. Every place where the flow is diverted can be likened either to the front (compressive) part of a venturi or to the rear (expansive) part. The path taken by any one molecule of the air is called a streamline. Upstream of our aeroplane the streamlines are drawn as straight parallel lines. At any point, if we take the streamline passing closest to the aircraft and project it downstream, then, if it intersects the aircraft, the flow is compressive. If it misses the aircraft (whose skin turns away from it) then the flow is expansive.

Let us start by looking at the flow over the very sharp leading edge of a thin wing. If the wing is really thin, like a safety-razor blade, the deflection of the air is almost zero. Instead of a strong shockwave, the leading edge creates a Mach wave of negligibly small amplitude, called the Mach line. Of course, this would really be a sheet or plane, but it is convenient to draw it two-dimensionally as a line. But a real wing has to have depth, so it is shown as a thin wedge, with leading-edge angle φ (Greek phi). This is naturally the angle through which the flow is diverted. When the air encounters such a real wing, a strong shockwave is caused, and this (as we saw previously) sets itself at a steeper angle than the Mach line. As the air passes through this shockwave its direction is changed to be parallel to the upper surface of the wedge. A physicist or aerodynamicist would disagree if I said the change in direction was instantaneous, but

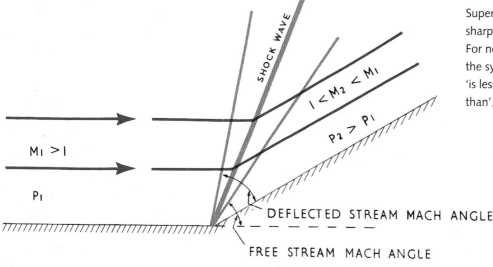

Supersonic flow round a sharp compressive corner. For non-mathematicians the symbols < and > mean 'is less than' and 'is greater than', respectively.

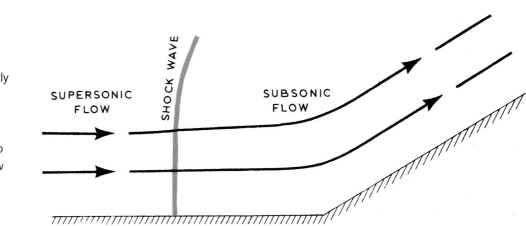

Flow round a compressive corner in which the deflection angle is sufficiently large and the Mach number sufficiently small for the velocity to be reduced below Mach 1.

the reader can think for himself how sudden it must be, bearing in mind that the shock is about one ten-thousandth of an inch thick and the flow is supersonic. Nothing like this can happen in subsonic airflow, where perfectly sharp corners are impossible.

As the flow passes through the shockwave its velocity is suddenly reduced. On the other hand, its pressure, density and temperature (and also, for technically minded readers, its entropy and refractive index) are just as suddenly increased. One can imagine a tiny region of the shockwave magnified thousands of times. This could be portrayed in terms of the streamlines of individual air molecules. Upstream their directed (supersonic) energy is high; downstream they form a dense jumble, their random (pressure and temperature) energy being greater than their directed energy. Incidentally, in this first example we have considered a sharp-edged aerofoil, but the result would have been just the same if it had been an outward corner somewhere along the side of the fuselage, diverting the flow by the same amount.

The greater the angle through which the flow is diverted – in other words, the greater the angle of a wedge or compressive corner – the stronger the shock. Increasing the shockwave's amplitude also increases its velocity of propagation, which causes it to adopt a progressively greater angle to the oncoming flow. This is because the stronger shock causes more intense heating, and this naturally increases the local speed of sound (because this is proportional to the square root of the absolute temperature). If we make the wedge angle φ larger and larger, or if we make the leading edge blunter, then, if the free-stream Mach number is only moderately supersonic, between 1.1 and 1.3 for example, we soon

reach a point where the forward speed of propagation of the strong shockwave is equal to that of the free stream. The slightest further increase in wedge angle or bluntness results in the shockwave ceasing to be attached. It will migrate forwards and form a detached, bow wave. The flow behind such a wave is subsonic. Directly in front of the leading edge such a wave is at 90° to the flow, and it is therefore called a normal shock (normal meaning at 90°). Any further increase in wedge angle or bluntness, or a reduction in free-stream Mach number, will cause this detached shock to move far ahead and disappear, as it ceases to be influenced by the solid body's presence.

Of course, we can study the same phenomenon in reverse by starting with our wedge or wing at rest and then increasing the Mach number. With a blunt leading edge, or a very large wedge angle φ, by the time the free-stream Mach number has reached about 0.8 the compressibility phenomena begin to happen. Can you imagine an infinitely weak normal shock being formed at an infinite distance? This is, of course, what happens in theory, but by the time the shock has travelled to within a few feet or metres of our wedge it has assumed a distinct form, with a very small area of supersonic flow in front of it and subsonic behind it. Further increase in M will drive the wave back to the leading edge. If the edge is sharp it will become attached; if blunt it will always stand off, though the distance will be small. In either event, further increase in M will cause the previously normal shock to lean backwards at the Mach angle.

In such a case, as for example ahead of the leading edge of a wing at Mach 2, the flow is supersonic everywhere. Clearly, around a real aircraft there are likely to be thousands of (mostly weak) shockwaves, each caused by some local compressive corner. In places these will interact, and they coalesce to form a single wave; no shockwave can ever pass through another. One can study what happens when a compressive corner is not a sharp angle but a smooth curve. At the very beginning of the curve (point A in the diagram) a weak Mach wave will form (AB). By the time the flow gets to C the Mach number has been slightly reduced and the flow deflected, so the wave formed here will be at a larger angle to the flow (CD). The portion OD is impossible, because CD cannot pass through AB; what actually happens is that the two waves merge into a single shock at an intermediate angle. When one considers all the numerous weak Mach waves coming from every point on the bend, it is possible to see that they all add together at some distance from the surface to form a single strong shockwave. Each weak Mach wave adds a little strength to the combined shock, and slightly increases its angle to the flow.

So far we have considered only compressive flow, but what about the other kind of supersonic flow? We saw earlier that, if we project the free-stream streamline that passes closest to the body and find that the surface of the body curves away from it, the flow is expansive. It is like that in the aft, enlarging, portion of the venturi: the flow speeds up, and its pressure and temperature are reduced (so too are density and refractive index, but entropy is unchanged). We can consider supersonic flow around either a sharp corner or a smoothly curved corner. Any aerodynamicist will appreciate that at subsonic speeds you simply cannot turn the airflow round a sharp-edged corner. The flow would break away, causing turbulence and high drag. Amazing as it may seem, at supersonic speed the airflow remains attached, and goes round the sharp bend perfectly. Starting at Mach 1, you can turn the flow through several sharp corners through a total angle of 129°! STOL (short take-off and landing) aircraft with high-lift flaps blown with air at over Mach 1 could generate fantastic lift! Expansions round corners, sharp or curved, are called Prandtl-Meyer expansions, because those great pioneers worked out the two-dimensional case in 1907–8, when the world air speed record was 39mph.

Consider the case where the corner is a smooth curve. We start with M greater than 1, and appropriate values of P, ρ (Greek rho, meaning density) and T. This air is unaware of the existence of the corner until it passes through the first weak Mach wave AB, which starts at the point where the aircraft skin first begins to curve

away. Instantaneously, P is reduced, and this both speeds up the flow (so that the next Mach wave leans back slightly more) and expands the flow so that its direction changes to remain exactly parallel to the skin. The air next passes through the second Mach wave CD (of course there are actually an infinite number of such waves, each coming from a different point on the curve, but we cannot draw them all). Wave CD again reduces the pressure and speeds up the flow, exactly enough to keep it flowing parallel to the boundary surface. Each wave leans back slightly more (precisely as much as the change in flow direction), so that eventually the airflow has come to the end of the expansive corner and carries on past the flat boundary surface downstream at its new, higher M and lower P, ρ and T. Note that the final Mach wave leans back at a significantly sharper angle than the first, showing the increase in velocity and Mach number (the opposite of what happens in subsonic expansion). Note also that, because of the expansion of the flow, and reduction in density, the streamlines downstream are further apart.

What happens if a supersonic flow encounters a sharp expansive corner? This is simply the limiting case of the smooth curve. We can imagine the radius of the curve progressively reduced, bringing all the sources of the Mach waves closer together but leaving the flow through each Mach wave unchanged. As the radius becomes zero (i.e. a sharp corner), all the Mach waves emanate from the same point. We still have the same Prandtl-Meyer expansion, which, adding all the millions of infinitely small changes in direction together, results in a smooth change in flow direction round the sharp corner. The final values of M, T, P and ρ will be exactly the same as in the flow round a gradual bend of the same total angle.

To anyone used to aircraft design this will probably seem amazing. With the possible exception of the Lockheed F-117A you simply do not make the air flow round sharp corners. The difference, of course, lies in the fundamentally opposite behaviour of air at subsonic and supersonic speed. At subsonic speed, which is what we are used to, air flowing round an expansive corner experiences a progressive reduction in velocity and increase in pressure. With care the designer can choose gently curving surfaces which, despite the adverse pressure gradient, do not cause breakaway of the flow. Any attempt at a sharp expansive curve causes such a strong adverse pressure gradient that the flow breaks away, causing chaotic turbulence and high drag. For

Supersonic flow round a compressive corner in the form of a smooth curve of significant radius (see text).

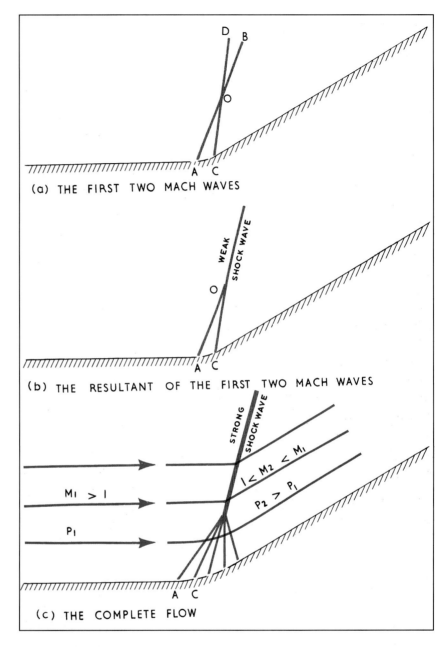

(a) THE FIRST TWO MACH WAVES

(b) THE RESULTANT OF THE FIRST TWO MACH WAVES

(c) THE COMPLETE FLOW

Below left: Supersonic flow round an expansive corner of significant radius

Below right: Supersonic flow round a sharp expansive corner.

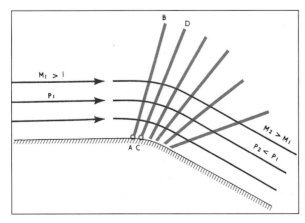

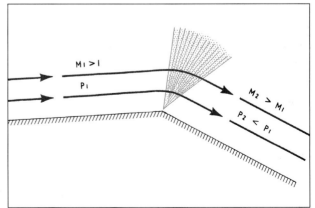

25

example, designers have to take great care to avoid flow breakaway around the rear underside of military cargo airlifters, where sharp expansive corners are difficult to avoid. If these aircraft could be supersonic there would be no problem. The flow can expand perfectly around the sharpest corner, because it follows the successive deflections of the Mach waves, and the boundary layer has no tendency to break away because the pressure gradient is favourable. If the pressure gradient is favourable – in other words if the air is always entering a region of lower pressure – there is no tendency for the flow to separate.

So why do we not see supersonic aircraft with sharp external expansive corners? In fact we do sometimes see such things – on the wings of missiles, as noted earlier. There are two common types of wing used on supersonic missiles. One is the double wedge, with a single sharp expansive corner along the line of maximum thickness. The other is the parallel double wedge, in which the leading and trailing wedges are separated by a central portion of constant thickness. There are two other common sections for purely supersonic surfaces. In the biconvex wing the upper and lower surfaces are smooth curves, sometimes

arcs of circles, joined at sharp (or small-radius) leading and trailing edges. The simplest profile of all is the single wedge, with a sharp leading edge and totally blunt squared-off trailing edge along the line of maximum thickness. This would be extremely inefficient at subsonic speeds, yet a 10 per cent single wedge was used for the vertical tail of the X-15 (Chapter 12). But I have not answered the question posed at the start of the paragraph. The fairly obvious answer is that manned aircraft cannot be designed purely on the basis of their supersonic behaviour. Even if they are carried aloft under a parent aircraft, they will eventually have to land, and this particularly demanding phase of flight has to be subsonic.

Having said that, it should already be apparent that you need one kind for subsonic flight and a totally different kind of wing for supersonic flight. A Nobel prize certainly awaits the aerodynamicist or engineer who can design a good supersonic wing that can also behave well (generate high lift coefficients) at typical aircraft take-off and landing speeds. So far we have good subsonic wings that are useless at over Mach 1, a few good supersonic wings that are useless at low speeds, and

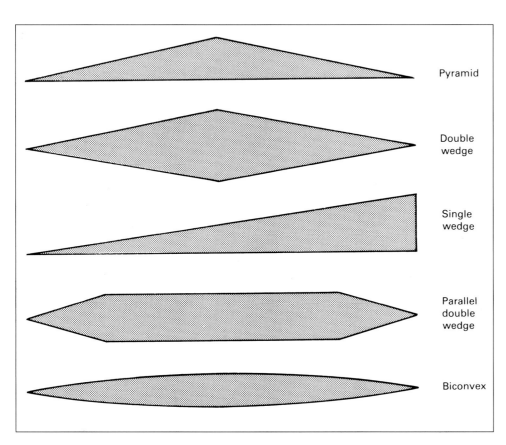

Pyramid

Double wedge

Single wedge

Parallel double wedge

Biconvex

Some supersonic aerofoil profiles (leading edge to the left). These are almost never used on manned aircraft.

a lot of intermediate wings that just manage to get by (rather poorly) in both regimes.

So the main thing remaining to be done in this chapter is to examine the behaviour of supersonic wings and what might be called 'ordinary' wings. Of the two, the supersonic wing is easier, because it simply consists of two things we have already studied: two compressive corners separated by an expansive corner. At low (subsonic) speeds the airflow will accelerate over the front half of the wing, reaching a maximum at the upper and lower points (actually lines, from root to tip) of maximum thickness. Here it will tend to break away, as we have seen, but we shall imagine that it stays attached, decelerating and increasing in pressure back to the trailing edge. As we increase the speed, eventually there comes a time where the local velocity at the two points of maximum thickness just reaches the speed of sound. A very small supersonic region is formed, with a Mach wave at its aft extremity. The free-stream Mach number where supersonic flow is first reached is called the critical Mach number, or M_{crit}. With aircraft such as the Concorde, the SR-71, or the MiG-25, this point was

of no more than passing interest, but 60 years ago, with such aircraft as the Gloster Meteor or Boeing B-47, it was of serious practical importance.

As we continue to increase the free-stream Mach number, the size of the supersonic region naturally increases. It always begins at the points above and below the wing where thickness is a maximum, but it spreads outwards, above and below, and rearwards, the upper and lower Mach waves becoming stronger and migrating to the trailing edge. When the free-stream Mach number is just below unity, something happens that we read about earlier: a second wave is formed far ahead of the wing. This second wave is at first a normal shock, at 90° to the flow, and it quickly comes back until (as our leading edge is sharp) it becomes attached. This occurs when the free-stream Mach number is just supersonic, in the region of 1.1. By this time the upper and lower aft normal shocks will have reached close to the trailing edge. Supersonic flow will prevail throughout, the last subsonic portion being between the front shock and the leading edge (this region would be larger and more important with a normal aeroplane wing with a rounded leading edge).

Transonic flow over a double-wedge wing at just above the critical Mach number. The broken line shows the size of the supersonic region.

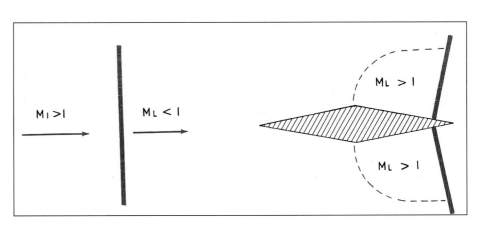

Flow past a double-wedge wing at just above the speed of sound (about M 1.02). The normal (front) shock is coming back to meet the leading edge.

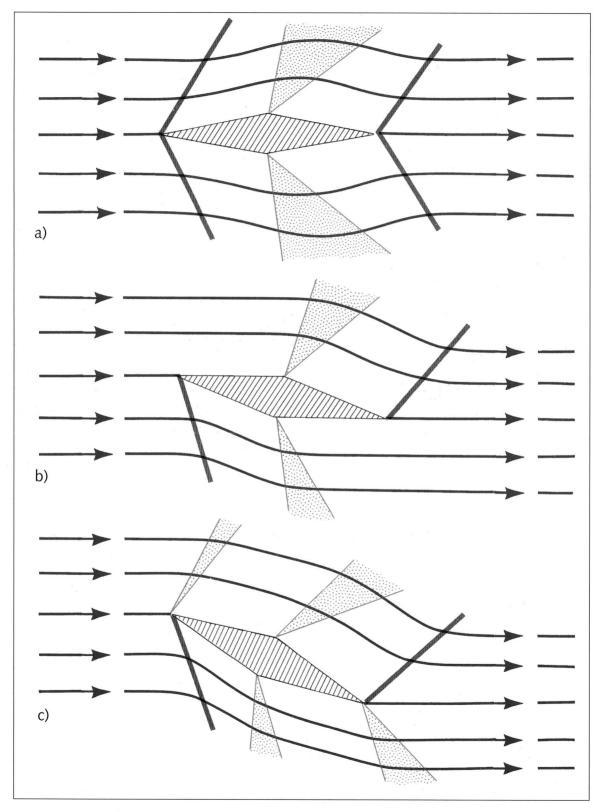

Supersonic flow past a double-wedge wing at three different values of (AOA) angle of attack: a) AOA zero; b) AOA equals wedge semi-angle; c) AOA greater than semi-angle.

It is clear that the transonic region is one of rapid and dramatic changes. With imperfect design a wing, especially when joined to a fuselage, engines and other parts, could easily suffer flow breakaway, with various kinds of instability and large changes in pitching moment, either nose-up or nose-down. Once the flow is supersonic everywhere, then, provided the aircraft is properly designed, there should be fewer problems. Three further diagrams show the flow at about M 1.2. By this time the supersonic flow is fully developed, with oblique shocks attached above and below the leading edge, Prandtl-Meyer expansions above and below the lines of maximum thickness, and oblique shocks above and below the trailing edge. At zero AOA (angle of attack) this symmetric wing would give no lift. Rotating it to an AOA equal to the semi-wedge angle would give positive lift, and eliminate the upper leading and lower trailing shocks, while strengthening the others. Further increase in AOA would turn the leading and trailing edges into expansive corners, generating increasingly powerful Prandtl-Meyer expansions where there were previously compressive shocks. The velocities, pressures and other variables bear no relationship whatever to what happens at subsonic speeds.

So much for a wing designed expressly for supersonic flight. With traditional wings the overall picture is rather similar, but with important differences. Conventional wings have rounded leading edges, so the front shockwave can never be attached. Even at Mach 6 it would still stand off a small distance (depending on leading-edge radius, the distance might be anything from a millimetre to an inch), with a subsonic region between the shock and the leading edge. Moreover, most wings are quite thick and have pronounced camber (curvature) over the upper surface, so M_{crit} might be as low as 0.5. I have no doubt that if the RAF Battle of Britain Memorial Flight dived their Lancaster vertically from its ceiling at full power they could get shockwaves along the region of maximum thickness, though I don't expect them to try!

The text below outlines how the flow pattern might develop as a conventional slow, Lancaster-type wing is accelerated to high subsonic speed (such a wing would

be difficult to force beyond Mach 1), while a group of seven drawings show the flow around a wing of 10 per cent t/c ratio, appreciably thicker than most modern fighter wings but still perfectly able to reach supersonic speeds.

With the old thick (18 per cent) wing we are likely to have a skin of thin metal, perhaps walked on by hobnailed boots, so the first tiny Mach waves could appear anywhere along the area of maximum curvature on the upper surface at about 22 per cent chord, at about M 0.62. In a real wing there will be many little wavelets, which will migrate to front and rear, possibly by several inches, only gradually taking form as a more definite Mach wave as the speed increases. With the profile chosen, this condition is reached by M 0.71, and at about the same point small Mach waves begin to appear on the undersurface. These coalesce and move to the rear, eventually catching up and stabilising opposite the upper wave. By this time, at around M 0.79, almost the whole flow over the wing's surface is supersonic. Any attempt to go faster would cause very high drag. With such wings the 'sound barrier' is very real.

With the 10 per cent symmetric profile shown in the drawings we can do much better. Whereas with the 18 per cent strongly cambered wing M_{crit} was about 0.62, with this thinner, less-cambered wing we do not reach local sonic speed until M 0.72, and, of course, the line of peak suction is much further aft. By M 0.85 the supersonic zone over the upper surface is quite large, and flow breakaway of the boundary layer, caused by the shockwave, has pushed a wedge of high pressure under

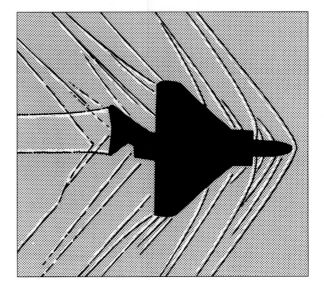

This Gloster Javelin is actually a tiny Dinky Toy, which was used in a wind tunnel to demonstrate shock formation at about Mach 1.4, a value never attained by the actual aircraft

Before transonic testing was possible in tunnels, designers often fired instrumented models of their aircraft with the aid of rockets. This model of the Convair F-102 failed to indicate that the original shape of this aircraft could not exceed Mach 1.

and horizontal tail cause the boundary layer to break away, leaving a turbulent wake. This wake can give rise to severe buffet, loss of lift (resulting in the concept of a 'shock stall', though since 1960 that expression has fallen into disuse) and, not least, loss of control effectiveness. But with today's wings and engine power we can simply push through it, and by Mach 0.95 almost the entire flow over the wing is supersonic. The shocks have migrated to the trailing edge, removing the turbulent wake to a location entirely downstream. At Mach 1.05 the bow shock has appeared, while the tail waves lean back ever more acutely. At 1.5 the aircraft is fully supersonic, except for the small region of subsonic flow ahead of the rounded leading edge. We shall never quite get rid of this, though, at Mach 2 and above, the bow wave is completely wrapped around it.

One extremely important fact, which, though discovered a century ago, is far from obvious, is that when a wing accelerates from subsonic to supersonic speed, the centre of pressure (CP) moves back by a large amount. The CP, which can also be called the centre of lift, is the point in the wing profile through which the resultant lift force acts. In a real wing, of course, it is a line running from the root to the tip on each side of the aircraft. At low speeds the CP is typically in the region of 25 per cent chord; in other words, it is about one-quarter of the way back from the leading edge. When completely supersonic flow is established, say at above Mach 1.2, the CP will be found to have settled down at about 50 per cent chord. This results in an extremely large nose-down pitching moment. Nothing like this happens in subsonic flight, and it is greatly accentuated by the fact that in supersonic aircraft the chord of the wing is far greater than in equivalent subsonic aircraft, so the change in longitudinal trim is proportionately much more powerful. The obvious way to counter this would be for the pilot to haul back on the stick while he winds on a lot of nose-up trim. This action would result in a considerable increase in drag, which is the last thing you want in supersonic

the foot of the original shock, to cause a bifurcated shock. This is also called a lambda shock, because it resembles the Greek letter λ. We need not bother about this, but almost all real shockwaves downstream of large supersonic regions have a split λ or bifurcated foot. With this wing 0.85 also brings the first region of supersonic flow on the undersurface.

This regime was the limit for the first generation of jet aircraft. The shockwaves above and below the wing

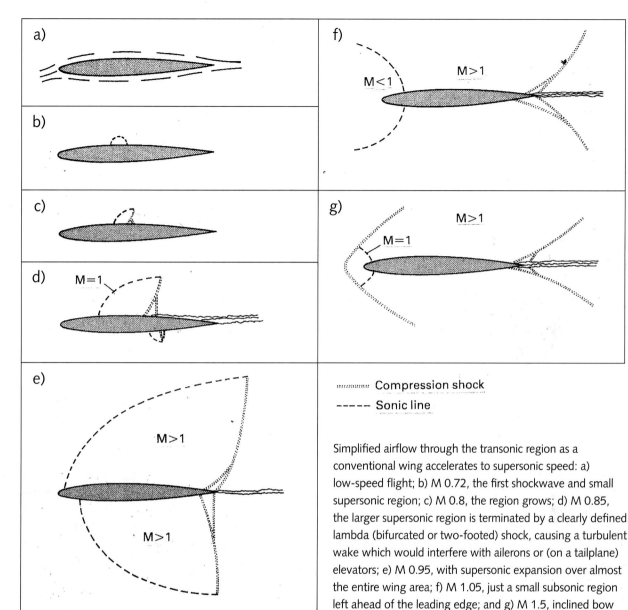

········· Compression shock

----- Sonic line

Simplified airflow through the transonic region as a conventional wing accelerates to supersonic speed: a) low-speed flight; b) M 0.72, the first shockwave and small supersonic region; c) M 0.8, the region grows; d) M 0.85, the larger supersonic region is terminated by a clearly defined lambda (bifurcated or two-footed) shock, causing a turbulent wake which would interfere with ailerons or (on a tailplane) elevators; e) M 0.95, with supersonic expansion over almost the entire wing area; f) M 1.05, just a small subsonic region left ahead of the leading edge; and g) M 1.5, inclined bow and tail waves and a very small subsonic region.

flight. A better answer is to shift the CG aft by pumping fuel to a special tank in the tail; on subsonic deceleration it is pumped back again. This ought to be a feature of all aircraft designed to cruise at supersonic speed.

What about drag? I commented earlier that, at Mach numbers anywhere near unity, a traditional supersonic wind tunnel is critically sensitive to any change in cross-section. Before 1950 most of them were extremely small, with working sections often measured in inches rather than feet, and though they could run at Mach 1 empty, if you put a tiny model aeroplane in the tunnel – even a model that would fit on a charm bracelet – its presence might instantly cause the tunnel to choke, with shockwaves from wall to wall. How galling it was, to find plots of CD rising steeply at Mach 0.7 to 0.8, only to vanish up into the unknown. Then they could be plotted again at Mach numbers beyond about 1.2. What everyone wanted to discover was what happened in between. This is not a book about research tools; suffice to say, with ventilated throats and variable flexible walls, tunnels were made by the 1950s that could explore the entire range of transonic Mach numbers.

Typical results are shown here. One set of curves gives an idea of how important it is to make supersonic wings

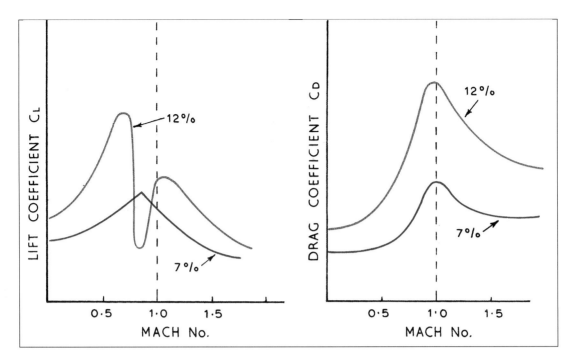

Generalised curves showing the effect of aerofoil thickness on lift and drag from subsonic to supersonic Mach numbers. In round figures, 12 per cent was typical of 1948 and 7 per cent was typical of 1953. The thinner the wing, the less the drag rise across Mach 1.

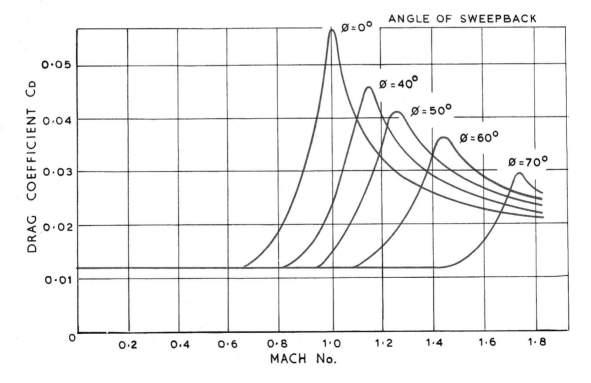

Generalised curves showing the variation of drag coefficient across the transonic region for a thin wing at different angles of sweepback. Note that greater sweepback is advantageous at lower Mach numbers but actually increases drag at higher supersonic speeds. In other words, provided the wing is thin enough, an unswept wing has lower drag at supersonic speeds.

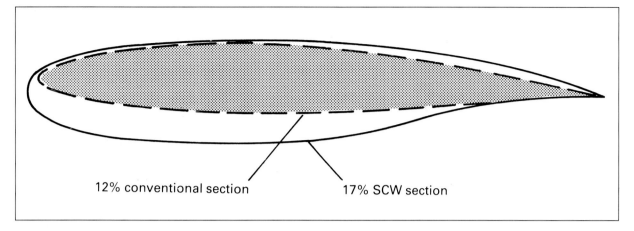

12% conventional section   17% SCW section

A supercritical wing superimposed on a thinner wing of conventional profile.

thin. By this we mean low t/c ratio. If you have enormous chord, as in Concorde or the XB-70, the actual thickness can be fully adequate for structural strength, or for the required internal fuel capacity. The other curves show the advantage of sweepback (or, now that we can make such wings, sweep-forward) in postponing transonic drag rise. But these curves also show that, while sweepback is excellent for an aircraft with a Mach limit of about 1.3, it is no advantage to the true supercruise (supersonic cruise) aircraft; indeed, it is positively disadvantageous. I shall go into much more detail about sweepback in later chapters, where we shall also look at one attempt to get the best of all worlds by pivoting the wings, and swinging them to any desired sweep angle.

There is a lot of fashion in aviation. Whereas in 1965 variable-geometry (VG) 'swing wings' were all the rage, by 1975 they were out of fashion. Some designers said the right sort of VG is to alter the profile, as for example in the F-16, where the entire leading and trailing edges are arranged automatically to pivot downwards (or, in the case of the trailing edge, also upwards) to match the unswept and very thin wing to various flight conditions. The one thing nobody argues about is that, for supersonic flight, you need a thin wing (and, of course, that goes for tail surfaces also, the vertical tail of the X-15 being a special case). Other things being equal, you can have a thicker wing if you sweep it acutely, and the highest t/c ratios of all are likely to be on VG swing-

wing aircraft. In general, the thinner the wing, the lower the supersonic (and probably the subsonic) CD, and the narrower the band of M over which tricky transonic phenomena are experienced. M_{crit} is raised, and this delay in the appearance of shockwaves to higher Mach numbers means that they will be weaker, and thus cause less wave drag. A further important point is that, on a thin wing, both the upper and lower shock-waves quickly migrate back to the trailing edge, rapidly sweeping the turbulent separated boundary layer back off the wing. The drawbacks of thin wings include the fact that the structure has to be very heavy in order to have adequate strength and rigidity, the maximum CL tends to be low, unless very clever high-lift systems are employed, and the leading edge needs to droop or be fitted with some form of flap or slat if it is to have satisfactory behaviour at higher AOA. A plain fixed leading edge, which of necessity has to be fairly sharp, stalls suddenly and with little warning at quite low AOA, resulting in extremely high landing speeds.

I ought to say a word about so-called 'supercritical' wings. These have no relevance to supersonic aircraft, but they do have something to do with shockwaves. Instead of having pronounced camber, with intense suction along the line of peak upper-surface acceleration, just behind the leading edge, the supercritical wing has a much flatter top, but quite a bulged underside; it also usually has a downturned trailing edge. Instead of giving intense lift over a narrow strip, as do traditional subsonic wings, the supercritical type (also called a roof-top or aft-loaded wing) generates a gentler lift over a very wide band across most of the upper surface. The result is either that the wing can fly at higher subsonic

Mach numbers or, more usually, that it can be made deeper, lighter and more capacious, and swept at a less-acute angle, all of which are highly advantageous. Such wings are quite unrelated to those of so-called supercruise fighters, such as the F-22A and Typhoon, which have thin profiles not very different from that of, say, an F-15.

Another major area of interest is propulsion. This is covered in most of the following chapters, especially Chapter 7. Not entirely facetiously, one could comment that, if you start with sufficient potential energy, you can go transonic with no propulsion at all. On 16 August 1960 Capt Joe Kittinger DFC, of the USAF, stepped out of a balloon at 102,200ft and fell 84,700ft before his main canopy was released. It was estimated that between 90,000ft and 60,000ft his speed was not less than 614mph (Mach 0.93), so the flow over major parts of his body would have been supersonic.

Chapter 4

THROUGH THE BARRIER

Any scientific research that is obviously unsullied by having anything to do with such lowly affairs as industry or commerce is called 'pure' or 'ivory tower' research. We could certainly apply these adjectives to the great Prandtl or Ackeret, who toiled away with small de Laval (expanding) nozzles to see what happened at Mach 1.5, while on the grass outside pioneer aviators with their caps back to front toiled equally hard just to get daylight under the wheels at 30mph. Even Busemann and Betz in the 1930s could not have known that their work would very quickly indeed become of the most vital practical importance.

Yet by December 1941 the first projects for swept-wing aircraft were already being considered by the DVL (German flight research centre) at Göttingen, the RLM (air ministry) and by the German aircraft industry. Such projects as the Focke-Wulf Ta 183, headed by Hans Multhopp (one of the few Germans to work in Britain postwar), the Messerschmitt P.1101 (led by Waldemar Voigt, who went to Wright Field) and the DFS 346 (under Felix Kracht, later the first Production Director of Airbus) all had their origins in 1942. These were the first aircraft projects in the world to be designed on the basis of the lately acquired knowledge of transonic airflow. In passing, it should be emphasised that the Me 163 rocket interceptor, and the Me 262 twin-jet fighter, were *not*. Though they seemed to have slight wing sweepback, and are often described as such, this had nothing to do with any desire to increase M_{crit}; Messerschmitt told me so himself. There was one more significant bit of foresight in 1942: Professor Alexander Lippisch, of the DFS (German gliding institute), who played the central role in the Me 163, filed a patent for the invention of the pivoted 'swing wing'.

At the same time, though the Me 163B *Komet* and Me 262A-1a *Schwalbe* and A-2a *Sturmvogel* did not incorporate any high-Mach knowledge in their design, they certainly deserve a mention in this book because they were the fastest aircraft of the Second World War. On 2 October 1941 the fourth prototype Me 163A reached 1,004km/h (623.87mph, though such accuracy of conversion is unjustified) at 4km (about 13,100ft), equivalent to Mach 0.84. Turbulence was encountered, immediately followed by an uncontrollable dive, and this led to a redesign of the wing. On 6 July 1944 the 12th prototype Me 262, fitted with a shallow low-drag canopy, reached exactly the same speed. The height was in the neighbourhood of 30,000ft, so in this case the Mach number was about 0.92, precisely the same as the figure reached by Martindale's Spitfire – the difference being that the Me 262 was in almost level flight. I am confident nothing else came near Mach 0.9 during the War.

I have always been surprised that, finding that tunnels were unable to explore the crucial transonic region, our aerodynamicists did not carry out free-flight experiments with high-speed aircraft deliberately modified to cause sharply accelerated local airflow. The only work of this kind of which I am aware concerned a Martin Baltimore bomber. By no means a particularly fast aircraft, it was found that the air over the top of the fuselage accelerated to approximately double the true airspeed, and one of these aircraft was used (possibly in dives) to test small wings and other specimens at Mach numbers around 0.9. Later the NACA fixed test specimens to the thickest part of the wings of two P-51 Mustangs.

It seems strange that no suitable aircraft (such as a Spitfire) should have been carefully modified for high-speed diving tests while carrying a test specimen mounted on a bulge causing local acceleration. I have no doubt

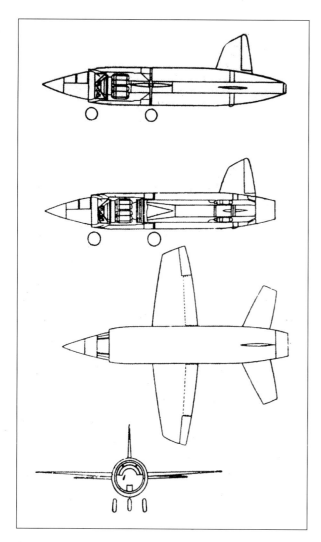

Nobody in aviation has ever had longer-range vision than Sir Frank Whittle. As I mentioned in the Introduction, he saw almost 80 years ago that, provided the propulsion system did not use a propeller, and provided designers solved possible problems of stability and control, aeroplanes could be designed to fly at high altitudes faster than sound. It was to enable the performance of aircraft to be multiplied that he invented the turbojet. Subsequently, in 1933, his calculations showed that, for supersonic speed to be reached, the jet velocity would have to be increased, so he invented afterburning (British term, reheat). As its name implies, this involves the burning of additional fuel in the jetpipe downstream of the turbine. With no highly stressed turbine blades to worry about, the temperature of the gas can be raised to white heat, typically 1,750°C, and with a suitably profiled nozzle the jet issues at supersonic speed. In the first supersonic turbojets the nozzle posed difficult problems. Not only does it have to retain strength at such extreme temperatures, but it has to change its size and profile. In normal subsonic operation it is small and convergent, whereas in full afterburner operation it is much larger and ends in a long divergent portion. Flow Mach number is 1 at the throat, the jet thereafter expanding and accelerating to several times the speed of sound in the divergent portion.

Whittle's combination of the turbojet plus afterburner obviously opened the way to the previously distant dream of supersonic flight. Of course, there was also another possible type of propulsion: the rocket. If this burns a solid propellant, its effective thrust is unlikely to last longer than a few seconds. Even if it is fed with liquid propellants, typically a fuel plus an oxidiser, these are consumed at such a high rate that there seemed to be no way of giving a rocket aeroplane an endurance under full power of more than two or three minutes. As supersonic flight could not be attempted except in the thin air at high altitudes, the only way to use rocket propulsion appeared to be to carry the aircraft up to high altitude by a large parent aircraft. An afterburning turbojet, on the other hand,

that Mach numbers up to about 1.2 would have been repeatedly attainable at very modest cost. It also seems odd that, until 1943, nobody in the world had thought it worth building an aeroplane specifically for high-speed research. Aircraft had been built for record-breaking, examples being the Me 209 and Napier-Heston Racer, but these were not only most unsuitable research tools, but were never even considered as such. I may add that there was no shortage of far-sighted people trying to get something done.

Towards the end of the war things were indeed started. Three programmes were launched for aircraft to fly faster than sound, one British, one German and one American. It is typical of the way we do things in Britain that we started long before our rivals, had a distinctly superior design, and then, at the eleventh hour, cancelled it for ridiculous political reasons.

No shockwave can pass through another. This is a typical representation of the flow past two aerofoils in tandem, such as a wing and a tailplane.

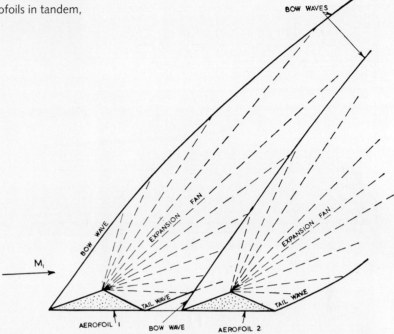

offered ample endurance for conventional take-off and landing.

In 1943, therefore, small groups of people in Germany and the United States were urging the development of supersonic aircraft, and news of the German work quickly reached London. For some reason, we got the idea the Germans were aiming at 1,000mph Though it seemed to many to be in the realm of pure fantasy, this report was judged serious enough to merit some action. In any case a supersonic aeroplane was an obviously sensible thing to develop; the timing was exactly right for it. And the remarkable thing is that, in the midst of a total war effort, one of the smallest firms in the British industry not only designed an outstanding aircraft but made much faster progress than their German and American rivals. Technical director George Miles said: 'Of course, if we'd been big we'd have been far slower.'

The Ministry of Supply experimental specification E.24/43 was one of the briefest ever issued. It merely requested an aeroplane that would take off and land in the normal manner, using a normal-sized aerodrome, and fly at 1,000mph at 36,000ft. On 8 October 1943 the contract for four M.52 aircraft was signed with tiny Miles Aircraft, near Reading, Berkshire. Quickly the first M.52 took shape, and the shape was that of a mid-winged projectile. Perhaps the only questionable feature

was that the cockpit was inside the glazed conical nose, which formed a centrebody in the air inlet to the engine. The complete nosecone could be separated in emergency, the ram pressure in the plenum chamber behind it then blowing it clear of the aircraft, when a stabilising drogue would be deployed. The tricycle landing gears retracted into the drum-like fuselage. The beautiful wing had a biconvex sharp-edged section, 7.5 per cent thick at the root and 4 per cent at the tip. From 11 August 1944 a Miles Falcon light aircraft was used to test an exact wooden replica of this wing, and later the one-piece 'slab' tailplanes.

Everything behaved as predicted, and lateral control remained beyond the stall. One can hardly imagine more cost-effective research, because even by 1944 standards the cost was insignificant. Miles' chief test pilot, Ken Waller, was to begin flying the M.52 itself from Boscombe Down in January 1947, on the 2,500lb thrust of a Power Jets W.2/700 turbojet with an 18.5in nozzle (just enough for safe flight). He was then to hand over to Cdr E.M. 'Winkle' Brown RN, who would do the flying with the No. 4 aft-fan augmentor added, with reheat both ahead of and behind the aft fan, and a variable nozzle ending with a diameter of 37in. The complete propulsion system was tested at sea level at 7,000lb thrust. At 36,000ft the thrust was calculated to

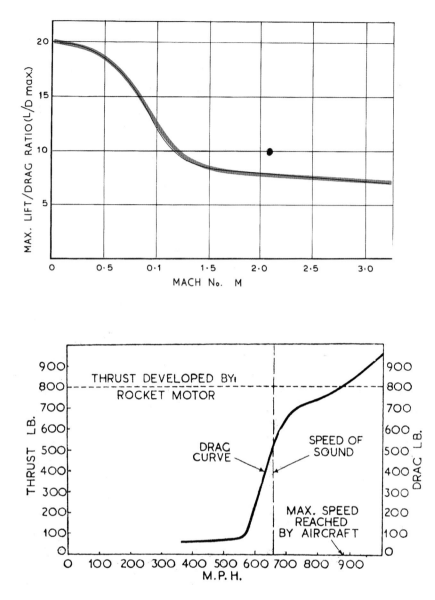

L/D is the basic measure of an aeroplane's aerodynamic efficiency. This shows how the best attainable L/D looked when Concorde was designed; the black dot shows where it would be possible to design a future SST

An aeroplane's maximum speed is reached when its drag in level flight equals its maximum thrust. This is an actual plot from one of the 1948 Vickers models. As it was shaped like an M.52 it suggests that that aeroplane ought to have been capable of diving beyond its 1,000mph design speed.

be 1,620lb at 500mph, 4,100lb at 1,000mph, and just over 6,000lb at 1,500mph

Even assuming that the ratio of lift to drag (L/D) was no better than 4, this would be enough for 1,000mph to be reached with an aircraft weighing 16,400lb. In fact the final estimates of L/D for the M.52 were about 7.7, and gross weight only 8,200lb. Looking back at the available information, my own belief is that the M.52 had the potential of reaching at least 1,500mph Subsequent events were to show that there was nothing wrong with the overall configuration, or the flight control system.

Tragically, in late 1945 the fact that vast amounts of German research material seemed to centre on the use

of sweepback for transonic aircraft made some of the top people in the Ministry of Supply think that the M.52 must be a complete mistake. I still cannot comprehend how it was that none of the aerodynamicists explained to them that sweepback is useful only in postponing the drag rise in the transonic regime. If you are going to fly at highly supersonic speeds then you forget about 'the barrier' and settle down in the regime on the other side, and sweepback is meaningless. In fact, as a diagram in Chapter 3 shows, provided it has a thin supersonic profile, in the supersonic regime the unswept wing has the lowest drag. The sharper the sweepback, the higher the drag at Mach numbers well beyond 1.

Such things were known in the Second World War, in Britain as well as in Germany. Yet the absence of sweep certainly played a major part in the cancellation of the M.52.

Ministry department heads were asked for their views on the programme, and their replies showed almost total ignorance of the difference between a transonic wing and a supersonic one. Added to this was an urgent need for economy, plus the personal view of the Director-General of Scientific Research, Ben Lockspeiser, that aeroplanes would not exceed the speed of sound for many years, if they ever did. My father told me that his boss, the Minister of Supply (then John Wilmot), had told him that the general view was that it would be 'foolish to persist with the ill-conceived M.52', and that it would be better to accept a proposal from Barnes Wallis at Vickers to carry out high-speed investigations using air-dropped models.

On 28 January 1946 Lockspeiser wrote to F. G. Miles, Managing Director and brother of George, curtly stating that the M.52 contract was cancelled. All parts, jigs and tooling were to be destroyed, and all data handed to the Ministry for transmission not only to Vickers-Armstrong but also to Bell Aircraft in the USA. Miles were specifically forbidden to say a word about this in public. Nobody was told anything until, at an Open Day at the RAE at Farnborough on 30 June 1946, a model of the M.52 was just visible tucked away in a corner. Pride of place was given to the Vickers model, which – reported *Flight* on 11 July – 'is based on the Miles M.52…'. This was the first time anyone had publicly referred to the E.24/43 programme.

A week later, on 18 July 1946, the Ministry held a press conference on high-speed flight. The newly knighted Sir Ben Lockspeiser announced: 'The impression that supersonic aircraft are just around the corner is quite erroneous, but the difficulties will be tackled by the use of rocket-driven models. We have not the heart to ask pilots to fly the high-speed models, so we shall make them radio-controlled.' He did not explain where he had heard or read this supposed erroneous impression. He made no mention of the M.52 until he was specifically asked, when he curtly said it had been cancelled 'for reasons of economy'. Subsequently, the Ministry took great pains to try to make the media forget about the M.52, repeatedly calling it 'a piece of dead research'. I tried to discover if it had become 'dead' through having unswept wings, and the fullest reply I ever got was that the 'The Ministry are not prepared to comment'. When my father asked about it he was warned that he should forget it.

What made it even stranger was that the first batch of models precisely repeated the configuration of the M.52. For two years the models, each dropped from a Mosquito and lost in the sea after its one flight, totally failed to work properly. At last, on 10 October 1948, a model actually worked, reaching Mach 1.38. The models had cost just over ten times as much as the M.52, and served merely to show that the M.52 had been a completely sound design. In February 1955 the Conservative government published a White Paper, *The Supply of Military Aircraft*, in which it was stated that cancellation of the M.52 'seriously delayed the progress of aeronautical research in the UK'. In 1970 a detailed study by Rolls-Royce at Bristol concluded: 'Re-examination of the M.52 in the light of present-day knowledge has shown that it … was capable of meeting the specification requirements.' Foolishly, when writing the first edition of this book I wrote to Sir Ben, asking how he viewed the decisions taken. He replied 'Old men forget…'.

In 1947 two of the small group of German aerodynamicists at Farnborough, Drs Multhopp and Winter, were given the job of carrying out the project design of a manned supersonic aircraft. It was a pedestrian effort, powered by a Rolls-Royce AJ.65 (later named Avon) turbojet without afterburner, with highly swept wings and a T-tail. The pilot lay prone in the nose inlet, and the landing was on retractable skids. Design Mach number was 1.24. Fortunately, no attempt was made to build it.

So what was the German project that was so quickly reported to Britain? It was described by Intelligence as 'a 1,000mph aircraft', though of course the Germans do not work in miles. I can only think it must have been the three-phase programme at the DFS, beginning with testing pressure cabins on the high-altitude DFS 228, then testing various supersonic wings on Heinkel P.1068 aircraft, and finally leading to the DFS 346. This was to be a pure supersonic research aircraft, designed under the leadership of Lippisch and Kracht in 1943–5. In fact, it was never completed, because its detailed design was to rest on the results obtained with the different configurations of P.1068, and these never flew either, though a great deal of model testing was done in tunnels. It has been reported that Siebel Flugzeugwerke were to build the two flight aircraft, and received the drawings in November 1944, but these drawings were little more

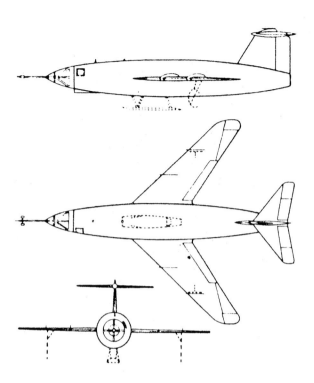

Most three-views of the DFS 346 have been inaccurate. This one appeared in the 1980s in Soviet and Polish journals.

The 346P glider. *Russian Aviation Research Trust*

than three-view general arrangements, and various outlines of particular parts. These were insufficient for construction to begin.

These drawings showed an aircraft designed to be carried on top of an unstated carrier aircraft to 10km (32,800ft) before release. It was then to fly under rocket power, followed by a long glide and landing on

a retractable central skid. The engines were to be two superimposed Walter HWK 509B rockets, burning the same propellant mix (called C-Stoff and T-Stoff) as in the Me 163B. Each engine would have had a sea-level thrust of 4,409lb, rising to about 5,000lb at high altitude, because rocket engines are not dependent on atmospheric oxygen and the propulsive jet reaches a higher velocity in very rarefied atmosphere. The wing was to be 9.2/7.5 per cent thick, swept at 45° and in the mid-position. The fuselage resembled a cigar, the pressurised cockpit occupying the glass nose, which was jettisonable in an emergency. The ailerons and the high-mounted elevators were divided into inner and outer sections, only the latter being designed for use at supersonic speeds. Control moments and even the wing-pitching moment were to be continuously recorded. The need for measuring the wing moment resulted from the expected large rearwards migration of the centre of pressure during the transonic acceleration, and the later return to the subsonic position.

The DFS 346 was designed to reach 2,765km/h (Mach 2.65, 1,718mph) at its ceiling of 35km (114,830ft). After the war the Soviet Union naturally took an interest in this work, and on 22 October 1946 formed OKB-2 (experimental construction bureau No. 2) at Podberezye, staffed mainly with former German researchers from the DFS, DVL and Siebel. Working initially under some duress, they completed the design of a succession of Type 346 aircraft, and in parallel managed mainly German workers who built them. One of the three Boeing B-29

bombers to land in the USSR, former USAAF No. 42-6256, was modified as the carrier, the attachment being under the bomber's starboard wing. In 1948 former Siebel test pilot Wolfgang Ziese made many test flights with the 346P glider. On 30 September 1949 he made the first with the 346-2, powered by a modified 109–509C rocket engine. Ziese was injured on landing (the skid failed to extend), but he shared many further flights with P.I. Kasmin of the LII (Soviet flight research institute) before having to bale out of the 346-3 on 14 September 1951. The programme was then terminated, without having reached Mach 2.

Though it had no direct connection with supersonic flight, the Messerschmitt P.1101 deserves mention as the first variable-sweep aircraft. It was eventually developed into the Bell X-5, as related in Chapter 8.

Before leaving consideration of German work, it is worth taking a brief look at the A.4 (the so-called 'V-2')

The 346-2 attached to the B-29. *Russian Aviation Research Trust*

rocket. This made its first successful flight in 1942. Before the A.4, not even the aerodynamics of bullets or shells had been properly studied, because it was so difficult to do. In fact, no man-made vehicle had ever propelled itself faster than the speed of sound. The A.4 was a ballistic vehicle, with no wings. It took off vertically, standing 46ft high, with a diameter of 5ft 6in and with four tail fins with a span of 11ft 9in. Fully fuelled it weighed, on average, 28,373lb. At lift-off the rocket engine, burning liquid oxygen and alcohol fed by turbopumps, imparted a thrust of about 26 tons, which steadily increased as the rocket climbed into thinner atmosphere. The trajectory was guided by aerodynamic rudders, and by small graphite thrust-vector vanes in the 3,000°C jet. The maximum speed reached was about 3,600mph, or Mach 5.

The enormity of this technical advance is obvious, though in 1941, when the rocket was being designed, the state of the art in accelerometers and gyroscopes was such that this otherwise deeply impressive delivery system was useless, unless the target was as big as London. As a teenager I was often awakened in the night by a heavy explosion, followed by the thunder-like rumble of the rocket's flight. You could hear the rumble for about 20 seconds, in other words, during the rocket's final 20 miles of travel, beginning at a height near to 100,000ft, where the shockwave energy would be very small. The quality of the sound was indistinguishable from thunder.

This may be a good place at which to bring in the de Havilland D.H.108. This was one of the most cost-effective of all research aircraft, and a brilliant example of the D.H. company's flair for doing bold things quickly. It was a tailless machine, reminiscent of the Me 163B but powered by a turbojet and with a genuine swept wing, the angle being no less than 43°. The central nacelle was based on that of the production Vampire fighter, with the same 3,000lb D.H. Goblin 2 turbojet and wing-root inlets. A swept fin and rudder were fitted,

The third de Havilland D.H.108 was the best looking and also the fastest. This was another British programme in which a lot of pioneer flying was accomplished on a trivial budget.

and the wings were equipped with plain flaps, leading-edge slats (fixed open on the first aircraft) and manual elevons, the latter giving all control in pitch and roll. The first D.H.108 made its first flight at Woodbridge on 15 May 1946, becoming the first true swept-wing jet aircraft to fly in the world.

Despite dire predictions of Dutch roll (uncommanded yaw/roll oscillations) and other problems by the Royal Aircraft Establishment, the first D.H.108 proved to have excellent handling at low speed, and also confirmed that the D.H.106 Comet airliner should have a basically unswept wing and a horizontal tail. The second D.H.108 was cleared for high-speed research, and had a less-blunt nose and a wing swept at 45°, with automatic slats. On 27 September 1946 it disintegrated at high speed and at low level over the Thames estuary, killing chief test pilot Geoffrey de Havilland Jr. I could not follow the eventual report's conclusions, because on the one hand it suggested that there had been sudden severe nose-down pitch, breaking off the wings downwards, while on the other it indicated 'catastrophic shock stall' leading to the elevons becoming ineffective. Such a shock stall on the outer wings would have made the aircraft pitch *nose-up*. Be that as it may, I was also puzzled by the statement that the break-up occurred at low level, because the figures quoted were 579mph and M 0.875, corresponding to about 35,000ft.

Naturally, de Havilland and the Ministry of Supply wished to press on, and on 24 July 1947 new chief test pilot John Cunningham flew the third D.H.108, built under a new contract. This had a more streamlined fuselage, power-boosted elevons, a more powerful Goblin engine, an ejection seat and, most importantly, trim flaps under the wings ahead of the flaps. These operated exactly like those of the P-38 Lightning and P-47 Thunderbolt. On 6 September 1948 John Derry, who had previously set an impressive 100km closed-circuit record in this low-powered aircraft at 605.23mph, started a 30° dive at 45,000ft with the intention of reaching Mach 0.96. The subsequent progress downhill was hair-raising. Derry had little control over the proceedings, the dive angle at times being over the vertical and the indicated Mach number reaching 1.04. Nothing had any effect until Derry switched the trim flaps open. There was an immediate gradual recovery, and the aircraft eventually regained level flight at 23,500ft at Mach 0.94. I cannot explain why this was said to have also been at an IAS of 'just under 500mph', because such an IAS would have equated with a TAS of some 723mph, clearly still supersonic.

Subsequently the third D.H.108 did much valuable flying, apparently exceeding Mach 1 again on 1 March 1949, when it entered an uncontrollable dive which ended with such sustained negative g that the aircraft completed the lower part of a bunt (outside loop). There was no structural failure, and the increasingly dense (i.e. warm) air progressively reduced the Mach number sufficiently for control to be restored at above 25,000ft. All three of these aircraft eventually crashed, killing their pilots, for totally unrelated reasons. Today we can see that perhaps a horizontal tail would have been a good thing to retain in an era of so many unknowns, but the D.H.108 deserves an honourable place in this book, because its skilled and courageous test pilots did a great deal to explore that notional barrier. On neither of the 'supersonic' dives was a sonic bang reported on the ground, but that is no real reason for doubting that Mach 1 was actually exceeded.

By this time, however, in the region of Los Angeles, sonic bangs had become commonplace, and they were being made by standard production F-86 fighters. Shortly afterwards, bangs were being made in the Soviet Union. But, as the world knows, the first man to accelerate under power to fly 'through the barrier' (and he could have been a 'Brit' named Waller or Brown) was an American, known to the world as 'Chuck' Yeager. This was because, in the United States, a supersonic research programme – with unswept wings – was not cancelled but allowed to bear fruit.

Yeager flew an aeroplane called the Bell XS-1; later the S (supersonic) was dropped, and I will call it X-1 from here on. It stemmed from pressure by Ezra Kotcher, a civilian engineer at the US Army's Wright Field. I never knew Kotcher. I am indebted for the information to Jay Miller's *The X-Planes* (Orion Books, 1988), where it is recorded that this far-sighted man had to wait from 1939 until 1944 before he could arouse interest in the notion of a special aeroplane to explore the transonic region. In 1944, a year after Miles began designing the M.52, things began to move, and the Wright Field proposal was sent to the NACA. There it was eagerly grasped by John Stack, then Assistant Director of the Langley Aeronautical Laboratory, who had been trying to get someone interested in the same idea since the mid-1930s. Nobody in the world did more to ensure that 'the sound barrier' would eventually be consigned to history. It is one of the ironies of fate that Stack should have been killed at an early age by being thrown from a horse.

The second Bell XS-1 (later X-1) parked on the apron at the Wheatfield, NY, plant. *PRM Aviation*

According to Miller's book, Kotcher had difficulty finding a contractor to build the proposed aircraft, until 30 November 1944. On that day Robert Woods, Chief of Preliminary Design at Bell Aircraft, at Buffalo, dropped in at Kotcher's office while on a visit to Wright Field. Kotcher worked the conversation round to the idea of a transonic aircraft, and found Woods enthusiastic. But Stack told me that Lawrence Bell had personally been drawing possible transonic aircraft since before the war, and that Woods had deliberately visited Kotcher to try to get something started. According to the NACA, they thought of the idea at Langley in 1943, started a big study programme and, in March 1944, organised a major seminar attended by the Army and Navy, where the NACA formally 'proposed that a jet-propelled airplane be built specifically for the purpose of flight research in the transonic region'. So the credit goes to the Army

at Wright Field, or maybe to Larry Bell, or maybe to NACA Langley. Who cares: all three played central roles in what happened afterwards. I should also mention that Theodor von Kármán, the Hungarian who became the greatest leader of aeronautical science in the United States, had been 'sold' on the possibility of supersonic flight ever since he attended the 1935 Volta congress, and put his enormous political weight behind all US efforts in this field.

After the March 1944 meeting the Navy and NACA got together on one programme, to which I will refer in the next chapter, while the Army, NACA and Bell went ahead on another. The Army and Bell signed the vital contract for research, design and manufacture of three X-1 aircraft on 16 March 1945. By this time almost all the major design decisions had been taken. These included: rocket rather than turbojet propulsion; thin but unswept wings rather than swept; a non-jettisonable cockpit or seat; and an air launch instead of conventional take-off. The latter was the last major decision, and it was rendered inevitable by long delays in the development of the turbopump to feed the propellants to the rocket engine. Changing to propellant feed by pressurised nitrogen cut the flight endurance under power so severely that a conventional take-off and climb to high altitude would have left inadequate propellants for transonic acceleration.

Construction of the three aircraft was assigned to Bell's plant at Wheatfield, NY. In most respects the X-1 was very simple, though its design g-limits of ±18 g were by any yardstick fantastic. Even today I do not know of any aircraft that comes near it, and I am sure that -18 would immediately break the pilot's seat harness and his neck with it. These unprecedented factors reflected the fact that the X-1 was designed to enter totally uncharted regions of flight, and at over Mach 1 even the thin air at high altitude can in a split second be lethal. In effect, Bell's team said: 'We don't know how to design the X-1, but we'll make sure it doesn't break.'

Except for the glazed cockpit roof (there was no windscreen as such), which had a steel frame with 20 panels of thick laminated glass, the entire airframe was aluminium alloy. The combination of immense strength and thin surfaces resulted in the first aircraft being completed with a wing 10 per cent thick, and a horizontal tail of 8 per cent. Flight controls were conventional and manual, but an innovation for the day was the pivoted tailplane, driven by an irreversible electric screwjack from +5° to -10° as a powerful trimmer. The small landing gears retracted into the fuselage under pressure from the main nitrogen bottles, which also assisted the bottled oxygen to keep the cockpit slightly pressurised.

The decision to use a rocket engine was a bold one. Outside Germany and the USSR nobody had much experience of this form of propulsion for manned aircraft. The small firm of Reaction Motors Inc. (RMI) had been started at Rockaway, on the shore of Long Island, NY, specifically for this requirement. Previously, the Aerojet Corporation at Azusa, California, had produced the XCAL-200 motor for the highly secret Northrop MX-334 flying wing, but this was a subsonic requirement calling for only 200lb thrust for a 3min 30sec burn time. The Bell transonic aircraft, which in US Army contract language was the MX-653, was going to need 30 times the thrust, and for a longer period under full power.

RMI chose to use the same propellants as in the German A.4, namely ethyl alcohol fuel and lox (liquid oxygen) as the oxidant (in the USA often called the oxidiser). It was originally intended that these should be fed under the required pressure of 315lb/sq in by a turbopump. Four separate individually controlled thrust

chambers were used, packed in a compact parallel group, with four triangulated steel tubes transmitting thrust to the fuselage. Each chamber was a cylinder of chrome-molybdenum steel, closed at the front and with the other end in the form of a convergent/divergent nozzle to accelerate the jet to a design figure of 6,182ft/sec at sea level. At the front were the various control valves and the BG ignition system which could fire each main chamber by feeding propellants to a small starting burner with an Eclipse booster coil at the closed head end. This avoided the very difficult task of developing an engine that was throttleable. As it was, the pilot could select 1,500, 3,000, 4,500 or 6,000lb thrust merely by using any desired number of chambers. In theory the best routine was to fire all four in quick succession, to get maximum acceleration.

Logically enough, this engine was called the RMI 6000C4, the military designation being XLR11-RM-3. At the time it was said to weigh only 210lb, but in fact a truer weight was 345lb, which in 1945 was still startlingly low for such power. Surprisingly, RMI said the thrust varied little with change in altitude. Of course, liquid oxygen was a new liquid to most people, and because of its need to be stored at -183°C (-298°F) the fuselage tank, holding 260gal, had to be thermally insulated to minimise boil-off. No provision was made for topping up the tank from the B-29 carrier aircraft, which carried the X-1 recessed under the fuselage.

Rather surprisingly painted a bright orange, the first X-1, Army serial 46-062, was secretly rolled out from Bell's Wheatfield factory on 27 December 1945. Strapped under a B-29, it was flown to Pinecastle AFB in Florida. From here, on 25 January 1946, Bell pilot Jack Woolams made the first gliding flight, after release from the B-29. He was positive about the handling, to the point of being ecstatic. He made nine further glides at Pinecastle. Then, on 11 October 1946, Chalmers 'Slick' Goodlin made the first flight of the No. 2 aircraft, 46-063, at Lake Muroc, California. This had the engine fitted, and on 9 December Goodlin made the first powered run, reaching Mach 0.75. On 10 April 1947 Goodlin began a fresh series of tests with No. 1, which had been to Bell's main Buffalo plant to have the wing replaced by a new one of only 8 per cent t/c ratio, and the horizontal tail replaced by one of 6 per cent.

On 6 August 1947, on Flight 38, the pilot was Capt Charles E. 'Chuck' Yeager, the first pilot nominated by the X-1's owners. He made three gliding flights to gain

The second Bell X-1 strapped (yes, with straps) under its carrier B-29 (45-21800). A special pit was dug in the South Base at Lake Muroc. The date was 9 December 1946, and Slick Goodlin was about to make the first powered flight. *PRM Aviation*

experience, and then on 29 August began powered missions. On 10 October, on flight 49, he got a Mach indication of 0.997. It is difficult to say what this meant, because there were still uncertainties about the pitot and static sources, and I don't believe you can read a Machmeter or printout that accurately. In any case, almost the entire flow around the aircraft must have been supersonic, the Mach number around the forward fuselage probably exceeding 1.3. At last, 14 October was set for the attempt to get a reading beyond the magic '1'. It says something about both Yeager and the whole Muroc operation that, on the previous night, Yeager suffered two broken ribs in a wild party, said nothing about it and boarded the X-1

next day! Levelling off at 40,000ft he fired the third of the four barrels. As before, he found the ride very rough and control virtually non-existent in the mid-0.9 region; once 0.98 was passed, the ride smoothed out and control was regained. Quickly the needle went to 1.02, waited several seconds and then jumped to 1.06. Yeager rightly put this down to rearward migration of the bow wave past the static source. Decelerating, there was 'a sharp turbulence bump' coming back through 0.98, and that was it.

So, on 14 October 1947, a man with two broken ribs riding a perfectly simple but very strong aeroplane with not a trace of sweepback pierced the dreaded barrier and came back to make a normal landing. Obviously, this was one of the greatest flights in the history of aviation, almost comparable with the pioneer flights of the Wrights in December 1903. From that day onwards, man has known that aeroplanes can be made to fly faster than sound. Yet, perversely, while Yeager's first penetration of the 'barrier' went almost unnoticed by the world's press – indeed, the Air Force tried to keep it secret, but *Aviation Week* coolly broke the story – seven years later the supposed sound barrier was still a popular subject for the media, especially in Europe. This was because – years later than in the USSR – prototype British and French fighters had taken to the air which could make supersonic bangs in a dive. Though understandable, the 1952–5 press interest was excessive. First, because hundreds of production F-86 Sabres had been making bangs since 1948, and second, because the whole idea of a barrier had become meaningless once designers learned how to make truly supersonic aircraft.

The X-1, the aircraft that conquered the barrier, went on to make 157 flights in its original form. Of these, 82 were made by the first aircraft, including the highest speed reached by an X-1, Mach 1.45, equating with 957mph at about 48,000ft, on 26 March 1948. On 5 January 1949 the same aircraft also made the only conventional take-off from the ground of any aircraft of the X-1 series. The No. 2 X-1, painted white instead of orange, and with generally better instrumentation, made 74 flights. The No. 3, 46-064, was by far the best. Again white, it had an XLR11-RM-5 engine whose propellants were fed by the long-awaited turbopump. In the first two X-1s the propellants were pressure-fed by large nitrogen bottles. In the third aircraft, even though a 26-gallon tank of concentrated hydrogen peroxide was added to drive the turbopump, elimination of the gas bottles enabled the main propellant tanks to be made larger. For example, the lox tank was increased in capacity from 260 gallons to 364. Moreover, the turbopump gave a sustained supply at full pressure for 4 minutes, instead of only 2.5 minutes. Though the original X-1 design probably could not have gone faster than Mach 1.45, the Dash-5 engine would have enabled trajectories to reach higher altitudes (the limit attained was 69,000ft) and sustain maximum speed for much longer.

Unfortunately the No. 3 aircraft made only one flight, No. 150 in the series, on 20 July 1951. On 9 November 1951, during ground tests before its second flight, it was completely destroyed by explosions and fire, which also engulfed its carrier aircraft, a Boeing EB-50A. Eventually the cause was traced to an instant chemical reaction between the lox and tricresyl phosphate used to impregnate ulmer-leather sealing gaskets.

When I read this I was stunned. At the very time the X-1s were being built, I was a vacation student at Westland Aircraft, where, among other things, I learned to weld. One welding method involves oxygen (used, for example, with gaseous acetylene). I was told: 'Never on any account allow a single drop of oil or organic material to get anywhere near the oxygen. It has been known for

100 years that that would cause a violent explosion.' So a galaxy of high-tech talent, working for the inspectors of the USAF, chose to mix lox with a material which not only causes instant explosions, but will detonate all by itself if frozen drops are struck with a small hammer. Ulmer leather was to cause the loss of *three further* Bell X-series aircraft, as we shall see.

Apart from this amazing example of incompatible materials, the overall X-1 programme enjoyed almost complete success, and a perhaps astonishing absence of errors or accidents. Considering that the basic design had been carried through in an environment of ignorance unknown in aviation since the turn of the twentieth century, this is truly remarkable. However, everything can be made better, and in November 1947 – just a month after the X-1 first exceeded Mach 1 – the newly created Air Force authorised the go-ahead on a second-generation aircraft, the Bell 58 or USAF MX-984. This could be designed with some assurance, not only of how to do it, but also of the performance to be expected. Every change was logical. The assured availability of a turbopump meant that the tankage could be enlarged (lox capacity, for example, was raised to 416 gallons), and this resulted in a longer and more tube-like fuselage. The cockpit was given a normal windscreen, and a fighter-type canopy hinged upward at the rear. Numerous detail changes were made to make the aircraft much easier to service and maintain.

There were intended to be four of the new aircraft. The X-1A (48-1384) was the baseline aircraft, with a 55in fuselage stretch, the new canopy, and many detail improvements. The X-1B (48-1385) was instrumented to investigate aerodynamic heating, as well as aerodynamic loads. The X-1C (48-1387) was to investigate supersonic gun armament, and to have yaw-damping fins above and below the outer wings and under the tail. The X-1D (48-1386) had a low-pressure turbopump feed, slightly greater propellant capacity, and modified instrumentation. In the event the X-1C was cancelled before the start of construction, and the first of the remaining three to be completed was the X-1D.

Early performance estimates for the X-1A/B/C/D indicated that, even using the same old Dash-5 rocket engine, it should be possible to attain Mach numbers and speeds roughly double those of the original X-1 series. This was very exciting, and because the leap in performance was so great I have deferred further discussion of these aircraft until Chapter 6, where I also describe the Navy/NACA effort that led to the Douglas D-558-II Skyrocket.

Following the dictum 'What goes up must come down', it was agreed from the outset that all of these supersonic research aircraft would have to make normal landings. A study was done to investigate the possibility of retrieval by hooking on beneath a carrier aircraft (probably not the one used for the launch), but it was pretty obvious that this idea was always a non-starter. So there was clearly going to be a need for a good airfield. Until the 1960s each major development in aircraft performance had to be paralleled by better airfields. At the start of the Second World War hundreds of pilots around the world were being killed each year trying to operate aircraft which landed at over 80mph from small rough fields that would just about have sufficed for a Tiger Moth. Clearly, the supersonic research aircraft were going to require landing fields not only at least 3 miles in length, but also with ample width so that the pilot did not have to follow the centreline of a narrow runway.

With commendable foresight we in Britain planned to build a super-wide runway 5 miles long in Bedfordshire, together with all necessary base infrastructure. This enormous test centre, initially for the Miles M.52, was to embrace three existing airfields, Thurleigh, Twinwood Farm and Little Staughton. The new postwar government quickly cancelled the whole scheme, leaving only Thurleigh to develop into the nation's chief aerodynamic research centre as RAE Bedford. My father showed me a memo saying (I don't remember the exact wording): 'Like the E.24/43 supersonic aircraft this grandiose scheme is a pointless waste of money'. So today, if a test pilot wants to land a really hot ship (dead-stick, damaged or with an incapacitated pilot) in Britain he has one choice, 06/24 at Boscombe, 10,537ft. In my view this would not be adequate for aircraft such as the X-1. Fortunately, in mid-California gigantic dry lakes provide an almost perfect site for today's Edwards AFB and NASA Dryden Flight Center, and any pilot will appreciate what an enormous (in both senses) national asset this is.

The only other matter to be covered here is the supersonic propeller. In the 15 years from 1945 the United States, mainly the Air Force and NACA, devoted prolonged efforts to seeing whether a propeller could be made to operate with reasonable efficiency at supersonic Mach numbers. Obviously, because the propeller rotates, the Mach number at any station along the blade is much higher than that of the aircraft. We have already seen how a Harvard trainer could attain a blade-tip Mach number just over unity at an aircraft Mach number of about 0.13. There are two ways of trying to make a propeller propel an aircraft at very high speed (say, above Mach 0.75). One answer is to use a fairly traditional type of propeller and cruise with a pitch angle so coarse that the rotational speed is extremely low, so that the blade Mach number is not much greater than that of the aircraft. This is obviously inefficient, because the lift force on every element along the blade acts not so much to propel the aircraft as to oppose rotation, thus demanding very great engine power. Despite this, this is the method used by the amazing Tu-20/142 'Bear' family of strategic platforms, capable of up to 588mph (Mach 0.87). Incidentally, their propellers make a distinctive growling snarl, audible for miles, just like ancient biplane bombers of the 1920s.

The alternative is to try to preserve blade angles of attack not much greater than those of ordinary propellers. This means that the bold decision must be taken to accept a Mach number well above that of the aircraft, even at the root of each blade, and to run the propeller at very high speed to maintain the efficient low angle of attack. Thus, we can make most of the lift force on each blade act not to oppose rotation, but to propel the aircraft. Such a propeller must obviously have extremely thin blades. It must also have a high hub/tip ratio; in other words the radius at the hub must be a relatively large fraction of the radius at the tip. There is less of a problem in designing a supersonic axial stage for a turbojet compressor, because here the hub/tip ratio is large in any case, there being short blades arranged around a large ring or disc. The first such transonic compressor was designed by Arkhip Lyul'ka in the Soviet Union for his AL-7 turbojet, first run in 1952. But with a propeller it is more difficult to arrange relatively short blades around a large hub, especially when their pitch must be variable.

Bearing in mind that propeller blades are miniature wings, then a transonic propeller needs sweptback or scimitar-shaped blades. A fully supersonic one needs even thinner blades, but they can be straight. All propellers, even those for Tiger Moths, suffer severe aerodynamic and centrifugal loads in flight, and for a supersonic propeller the stresses, and the possibility of severe turbulence and flutter, are much worse. From the aeroelastic and mechanical viewpoints straight blades are infinitely preferable to scimitars. Yet another intractable problem with supersonic propellers is that of noise. Can you imagine what a spiralling circular shockwave sounds like? Or a Harvard whose blades make the same noise not just at the tips but all the way to the roots?

The first supersonic propeller in the world to fly was a 10ft four-blader made by Curtiss Electric, driven by an Allison XT38 turboprop in the nose of the XF-88B. This aircraft was originally the first McDonnell XF-88 Voodoo prototype escort fighter, normally powered by two Westinghouse J34 turbojets. It first flew with the propeller on 14 April 1953, and despite many problems carried out a considerable amount of research from NACA Langley, including at least one shallow dive to beyond Mach 1. Various propellers were to be used, but the only known photographs show a unit with four thin rectangular blades, tapered in chord but thickened to a circular section at the root.

Next to fly, after prolonged delays, was the XF-84H. This was a Republic RF-84F Thunderflash with the J65

turbojet replaced by a 5,850ehp. Allison XT40 coupled turboprop, fed via the modified wing-root inlets and exhausting at the tail. Long shafting drove the main gearbox in the nose, coupled to the propeller under test. Because of the large diameter of the spinner, the hub/ tip ratio was high enough to get high Mach even at the roots. It was intended to test propellers by Aeroproducts, Hamilton Standard and Curtiss. In the event, only an Aeroproducts unit was tested, with three rectangular blades of 12ft diameter, maintaining their thinness right to the root. To take out some of the strong spiral twist in the slipstream a canted delta fin (called a 'vortex gate') was added behind the canopy, and a completely new T-tail was fitted. No performance figures were published,

Acute nausea was caused to people outside the XF-84H by the noise of its supersonic propeller running at high power. *PRM Aviation*

and only one of the three planned conversions was ever flown. Pilot Henry Beaird liked the aircraft, but the noise caused 'acute nausea' to everyone else.

So far as I am aware the only other supersonic propellers to have flown were two pairs of Curtiss Turboelectric units tested on the two Boeing XB-47Ds. These were previously B-47B Stratojet bombers, converted to flight test not only the propellers but also the Wright YT49 engine. This impressive single-shaft turboprop was virtually a J65 Sapphire plus a reduction gearbox, and a forward shaft drive. It was allegedly rated at 9,710ehp, but I doubt that this was achieved, even though the potential was even greater. A single YT49 replaced each inboard pair of J47 turbojets, and the propellers got a usefully large hub/tip ratio by being tailored to the outside diameter of the pod, the air inlet being through the hub. Each unit had four thin rectangular hollow-steel blades, with a diameter of 15ft. The first XB-47D made its first flight on 26 August 1955, and eventually a level speed of 597mph at 13,500ft was attained, an aircraft Mach number of 0.83, and a mean blade Mach number of 1.49.

I am not aware of any work on supersonic propellers having been done since (but see the *Montaigne Mach-Buster*, Chapter 11). Today's propfans are designed for tip Mach numbers from 0.7 to 1.1, but everything possible is done to minimise shockwave formation and noise, and in every case the aircraft concerned is subsonic.

Chapter 5

THE FIRST SUPERSONIC FIGHTERS

Looking back from a distance of some 65 years at large numbers of general-arrangement drawings of early jet fighters – most of which never got off the drawing board – one is inevitably struck by the fact that each falls into one of two clearly defined groups. About three-quarters look awful: they have the appearance of particularly bad piston-engined fighters which just happened to have jet engines. The rest look good, because they were created by teams that understood a little about high-speed aerodynamics.

Unquestionably the best of all this early crop of jet fighters was the XP-86 Sabre. In 1944 North American Aviation (NAA) began work on the NA-134 fighter for the US Navy and the NA-140 for the USAAF. The former entered service as the FJ-1 Fury, and it was a competent straight-wing machine. With the NA-140, however, the company showed commercial courage in deliberately delaying this design in order to incorporate the newly available German data on swept wings. From August 1945 NAA carried out more than 1,000 model tests, almost all on surfaces swept at 35°, and with about half the total concerned with leading-edge slats, which appeared to be necessary to confer adequate stability at high AOA at low speeds. The results were extremely encouraging. The wing finally adopted was aerodynamically almost exactly the same as one studied by Messerschmitt AG in spring 1942, when it was hoped to fit swept surfaces to the Me 262. The t/c ratio was 11 per cent at the root and 10 at the tip, and a crucial factor was that the flight controls were hydraulically powered.

The prototype XP-86 Sabre first flew on 1 October 1947. The engine was one of the first American axial turbojets, a General Electric TG-180, manufactured by Chevrolet as a J35-C-3. Its take-off thrust was 3,750lb, fully adequate for initial flying, and results were outstanding in all respects. Apart from changing to the more powerful General Electric J47 engine, the XP-86 needed very little change before it went into production as the F-86A. Handling at all speeds in all attitudes was superb, and arguably superior to that of any straight-wing fighter. On 26 April 1948 test pilot George Welch exceeded Mach 1 in a shallow dive, and from that time onwards many thousands of Sabres were to make sonic bangs all over the world. Recovery was straightforward, full control remaining throughout; there was no severe buffet or shock stall, and the concept of a sound barrier clearly did not apply to this aircraft. Had it had more power, of course, it could have gone supersonic on the level. Later versions had engines rated at up to 7,650lb thrust with afterburner, and (the J73, in the F-86H) 8,920lb without. During the Korean war, combats with the MiG-15, which in many ways had a similar performance apart from an inability to dive beyond Mach 0.91, led to the surprising decision to replace the automatic slats by a 'hard' leading edge of slightly extended chord. This took effect during production of the F-86 and F-25. Subsequently, slats of an improved design were restored to production, and retrofitted to most 'hard' aircraft. In no case was low-speed stability or handling seriously affected.

Although it was a subsonic aircraft, the F-86 deserves its place in this book because it introduced thousands of pilots to brief periods of transonic flight. It was such a splendid and popular aircraft that NAA made only two tandem 'two-stick' trainers, one on company charge and the other for the USAF. Instead, pilots posted to F-86 squadrons in 23 countries were simply allowed to go off and make sonic bangs. The F-86 also introduced airshow crowds to this new phenomenon. I shall never forget the 1951 Paris airshow where, amid hordes of 'plank-wing' aircraft, a Canadair-built Sabre took off and climbed

A classic design: the North American F-86 Sabre. *PRM Aviation*

to 40,000ft, a tiny speck precisely over Le Bourget. I watched the 410 Sqn RCAF pilot roll over on his back and pull through the second half of a loop, going back the way he had come. Almost half a minute later came the tremendous bang, as the shockwave reached ground level, and an almost hysterical commentator Jacques Noetinger shouted over the loudspeakers 'Il a franchi le mur du son!' And this was not only a regular squadron aircraft, with full armament, but one made under licence in a second country.

Britain's reluctance to build transonic fighters became almost absurd. In 1953 Hawker Aircraft was well advanced with the P.1083, a Hunter with a thin 50° wing and afterburning Avon. The Air Staff had this cancelled, and replaced by the P.1099 with the original wing and an Avon 200 without reheat, guaranteeing that the next

fighter, unlike the P.1083, would be subsonic. In the same way, Supermarine followed the Swift with the Type 545, a completely new fighter with a particularly good wing and an afterburning engine. It was also designed according to the area rule, which features later in this chapter. The 545 was proposed in two stages, the first with an Avon, to reach Mach. 1.3, and the second with an RB.106 to attain Mach 1.6. In 1955, when the first aircraft was structurally complete, the RAF cancelled it, to make sure it would have no supersonic fighter until 1960.

Things were different in the Soviet Union. In the late 1940s there was intense competition between the fighter OKBs (experimental construction bureaux) of Mikoyan/Guryevich (MiG) and Lavochkin. Both began with straight-wing aircraft powered by modified German axial turbojets, but the extraordinary decision of the British government to ship to Moscow the latest British turbojets enabled the rivals very quickly to produce much better fighters powered by Soviet pirated copies of the Rolls-Royce Derwent 5 and Nene. In September 1948 the La-176, powered by the RD-45

(Nene), boldly featured a wing swept not at the 35° that had become almost universal in other countries, but at 45°. Semyon Lavochkin's team had already flown the La-160 with a 35° wing of 9.5 per cent t/c ratio, as well as the structurally challenging La-174TK, TK meaning *Tonkoye Krylo* or thin wing. Its 6 per cent profile was the thinnest in the world for a further six years. The La-176 proved superior, and on 26 September 1948 Capt O. V. Sokolovskii took it beyond Mach 1 in a dive (according to the instrumentation). From early 1949 the La-176 routinely made sonic bangs.

Lavochkin was already working on the much heavier and more powerful La-190, with the big Lyul'ka AL-5 axial engine, and a remarkable wing combining 6.1 per cent thickness with 55° sweep. On 11 March 1951 the La-190 held Mach 1.03 in level flight, the speed being about 740mph at 16,400ft. Sadly for Lavochkin, it was destined to be yet another good prototype beaten by an even better one, in this case the MiG SM series. The MiG bureau had flown the SI, the first prototype of the MiG-17, in January 1950. Artyom Mikoyan announced that the SI had reached Mach 1.03 on the level at only 7,200ft on 21 February. Arguments then ensued over the instrumentation. The SI, also designated as the I-330 (fighter type 330), crashed a month later. Certainly, no other MiG-17 was supersonic on the level, but the claim

to have exceeded Mach 1 in such dense (and very cold) air is noteworthy. But the prolific MiG team was already far advanced with the M programme for a totally new fighter, designed to exceed Mach 1 in level flight. The first members of this family were the I-350, powered by a Lyul'ka AL-5, and the SM-2, with two of the slim AM-5 engines which were later developed into the Tumanskii RD-9B. Both aircraft had prominent T-tails, but what was more impressive was the wing.

In the period around 1950 designers were understandably daunted by the structural problems of very thin wings, which obviously have far less torsional rigidity than thick ones. One solution was to reduce the aspect ratio, making the wing broad, but of small span. Another was to adopt the delta shape, as described later. Another was to accept that the wing would twist, and put the ailerons inboard. Had more been known about them, a further choice would have been to replace the ailerons by spoilers, or to use a rolling tailplane (taileron). The MiG

This heavily retouched photograph is almost the only one available of the La-176, which could be described as the first supersonic fighter. *Russian Aviation Research Trust*

The MiG I-360/1, also designated as the SM-2/1. *Russian Aviation Research Trust*

team did none of these things, yet gave the SM-2 and its successors a beautifully slender wing, with quite a high aspect ratio, swept at 55° and with 8.24 per cent thickness. Boldly, they put conventional ailerons at the tips, high-lift flaps inboard, and left a plain fixed leading edge. It gave no trouble, though other parts did, notably the tail. Eventually this family was to spawn an extraordinary series of variations, but the main production aircraft were all different kinds of MiG-19. These had twin RD-9B afterburning engines, a plain nose inlet, and a horizontal tail which was first brought down to the fuselage, and then replaced by a one-piece 'slab' on each side, resulting in the designation MiG-19S (from *Stabilisator*).

The MiG-19 series had level speeds in the 1,432–1,582km/h (890–983mph) range, corresponding to Mach numbers of 1.35–1.45. Large numbers were built, production continuing in China (as the J-6) until 1983. Although it seldom hit the headlines, as did the MiG

fighters that preceded and followed it, the MiG-19 deserves a place in history as the first truly supersonic fighter to enter service. One group of prototypes, the SM-12 family, raised the speed to 1,930km/h (about Mach 1.8), chiefly by having a more efficient nose inlet with a centrebody. I was surprised that these never led to a production version.

One of the difficulties in deciding which was the first supersonic fighter, apart from the La-190 prototype, is that the initial production MiG-19 differed in major features from the I-350 flown in May 1951 and the SM-2 flown a year later. Another contender for the title is the American F-100 Super Sabre. Though this did not fly as a prototype until 25 May 1953, the impressive combination of technical capability and sheer aggressive drive by both the customer (the USAF) and the contractor led to the first production F-100A being completed as early as 28 September *of the same year*. Thus, thanks to the contrast in development pace between first flight and clearance for service, the F-100 can be argued to have been the first supersonic fighter to become operational. Unfortunately, it did so in a dangerously flawed condition.

In 1947, when the XP-86 Sabre began flying, the immediate lines of development were both based on a

very similar airframe. The NA-165 featured nose radar, and a chin inlet feeding a D-series J47 engine with afterburner, giving some 7,500lb thrust; this led to the production F-86D. The NA-157 had a completely new fuselage with flush NACA side inlets feeding a Pratt & Whitney J48 (licensed Rolls-Royce Tay) to give 8,300lb thrust with afterburner. But NAA's Vice-President Engineering Ray Rice soon came to the conclusion not only that a supersonic fighter was a practical proposition, but also that it would have to be a completely new design with a new and much more powerful engine. By the time the NA-180 design was completed in 1951, becoming the F-100, the only similarity with the F-86 was the general configuration.

A key factor was Pratt & Whitney's development of the JT3 turbojet, initially produced for the Air Force as the J57. Having cut their teeth on the British-derived J42 and J48 centrifugal turbojets, the Connecticut firm's superb design team embarked on an all-new axial engine. Its features included splitting the compressor into two parts, an LP (low-pressure) spool followed by an HP (high-pressure) section, each driven by its own turbine, the LP shaft running down the centre of the faster-rotating tubular HP shaft. This engine not only offered a thrust of 10,000lb dry and 15,000lb with full

Yet another classic design was the NA-180, the two prototypes of the F-100 Super Sabre. This is the F-100D fighter-bomber version that featured an increased root chord. *PRM Aviation*

(variable) afterburner, but its high pressure-ratio also promised outstanding fuel economy.

Today we would say it was huge and ponderous, occupying half the volume of the F-100's fuselage, but 55 years ago it was deeply impressive. It was fed by a straight-through duct from a nose inlet of distinctive shape and with the (then unfamiliar) sharp edges necessary for supersonic flight. The wing had to be a compromise, as in any fighter, but from the start it had 45° sweep and a t/c ratio of 6 per cent. The area finally decided upon was 385sq ft, compared with the F-86's 287.9, giving a clear indication that what was to become known (from their designation numbers) as the USAF's Century Series of supersonic fighters were going to be significantly larger and heavier than any seen previously. Despite the immense strength of the wing, NAA was anxious to avoid aileron reversal, caused by inadequate torsional stiffness,

and in the end put the ailerons inboard. As a result the F-100A had no flaps, merely slats along the leading edge, so that a squadron 'Hun' pilot was later to describe each landing as 'a sort of controlled crash'. The horizontal tail comprised one-piece slabs mounted at the bottom of the fuselage, and the flight-control power units were roughly four times as powerful as those of the F-86. Under the belly was a giant door-type speed brake, which ruled out the carriage of stores, except under the wings.

The YF-100 prototype exceeded Mach 1 on the level at 35,000ft on its first flight, winning two beers for chase pilot Col Frank 'Pete' Everest, Chief of Flight Test at Edwards. Everything went well and, under the newly formulated Cook-Craigie Plan, full-scale production followed right behind the two prototypes, the initial low rate progressively speeding up as hurdles were cleared. On 19 October 1953 chief test pilot George Welch demonstrated the first production F-100 A to the world's press, among other things flying past at treetop height at over Mach 1, the colossal bang being followed by the tinkling of broken airport windows. By this time production was beginning to roll. The F-100A differed only in details from the prototypes, the most obvious being a carefully calculated reduction in the height of the fin and rudder.

Everest appointed himself F-100 Project Pilot, and in exploring the boundaries of the flight envelope he and Capt Zeke Hopkins discovered a tendency towards roll-coupling (described below). They considered this to be potentially dangerous, even though they did not fully understand the problem. A huge political row ensued. Under the Cook-Craigie Plan the production rate was increasing sharply, and both NAA and the USAF had enormous vested interests in rejecting Everest's findings. In a difficult situation, test pilot Welch gave formal evidence that the F-100A complied with USAF specifications, and was 'one of the safest fighters'. So production continued to build up, and in September 1954 the 479th Fighter Wing was activated at George AFB, near Victorville, California. I visited them soon afterwards, and found the accent strongly on safety, because four 'Huns' had broken up at high altitude,

North American's F-100B became the F-107, notable for its dorsal inlet and one-piece powered tailplanes and rudder.

apparently because of some kind of control failure. The pilots did not survive, one of them being Air Cdre Geoffrey Stephenson, Commandant of the RAF Central Fighter Establishment, and a man of exceptional experience. On 12 October 1954 Welch took off in an F-100A to do the very last and toughest of the required structural flight tests, calling for a 7.5 g dive pull-out at the very limit of IAS and Mach number. The fighter broke into small fragments, and Welch was killed. An exhaustive investigation showed that the aircraft had disintegrated as the result of a violent yaw to the right.

Thus was the phenomenon of inertia coupling dramatically brought to the attention of the supersonic designers. Traditional aircraft tend to have short bodies and big wingspans, whereas with supersonic aircraft the reverse is true. The various masses tend to be distributed along a line from nose to tail, and any attempt to induce pitching or rolling results in centrifugal forces proportional to the square of the rate of rotation. With a big and long fuselage, and relatively small wing, the aircraft can very quickly become directionally unstable. This tendency is accentuated in a dive at high speed, with the principal roll axis inclined downwards (in this case at about 80°). Everest was completely vindicated, the F-100 was grounded, and 70 already delivered, plus 90 on the production line, were rebuilt with increased span and a much taller vertical tail.

The grounding did not apply to NAA test flying, and on 26 February 1955 test pilot George Smith unexpectedly found himself testing 53-1659. It was a Saturday, and he had only called in at the plant to finish writing a report. He did not bother to put on boots or a flight suit, but wore his parachute over a sports shirt and slacks. An obscure hydraulic fault resulted in the aircraft going into an ever-steeper dive. As it went beyond the vertical Smith ejected, noting the Mach reading of 1.05. His list of injuries was awesome; he 'approached close to the limit for human survival'. It was established that he had ejected at 780mph in the dense air at about 6,500ft. This was the first supersonic ejection, and a lot was learned from it.

Subsequently the 'Hun' had a particularly long and active career, proving in Vietnam its ability to fly both fighter and attack missions. A version designated F-100B was developed into the F-107, which a USAF general described as 'the best airplane we ever canceled'. Powered by a huge Pratt & Whitney J75 engine rated at 24,500lb, it was noteworthy for its visually striking dorsal inlet. Designing an inlet for a supersonic turbojet is far from simple, as explained in Chapter 7. The F-107 was also noteworthy for its one-piece 'slab' vertical tail, and for finally solving the problem of aileron reversal by not having any ailerons. It was one of the first aircraft to have roll control by spoilers.

The McDonnell XF-88 Voodoo, first flown in October 1948, was a superb design, potentially supersonic but crippled by inadequate engines. The XF-88 was thereupon enlarged to become the F-101, still named Voodoo. Powered by *two* J57s, it was in 1954, by a huge margin, the most powerful fighter then to have flown. It also had an exceptionally high wing loading. In 1939 concern was voiced over the wing loadings of the new monoplane fighters, the Bf 109 figure being as high as 26.5lb/sq ft. Designers at that time would have been incredulous had they been told that in 15 years the F-101's wings would reach 142lb/sq ft! Such a loading reflected the increasing power of afterburning turbojets, which could thrust fighters to high take-off speeds, and keep speed up in sustained hard manoeuvres. In fact, today's wing loadings are invariably lower. The F-22A Raptor, for example, has a maximum wing loading of 71.4lb/sq ft.

Convair's F-102 Delta Dagger was the classic case of a supersonic aircraft that proved incapable of exceeding the speed of sound. This was doubly odd, because it was the first fighter to feature the tailless delta configuration, which later worked perfectly well in other aircraft. The pioneer of this layout was the German Alex Lippisch, whom we met in Chapter 4. After the war he was taken to Washington, eventually becoming Director of the Collins Aeronautical Laboratory, at Cedar Rapids, Iowa. Before he went there, he had helped Convair to build a delta fighter for the Air Force, and Douglas to build a near-delta fighter for the Navy (described later in this chapter).

In its purest form a delta is a triangle, which is the shape of capital Delta, the fourth letter of the Greek alphabet. You can, of course, add a horizontal tail, as in the MiG-21, and this enables the wing to be fitted with high-lift flaps. But the pure delta is a tailless aircraft, the wing carrying elevons on its trailing edge for both pitch and roll. Deltas come in many forms, and in Chapter 10 the ogival or slender delta is important. Lippisch favoured the plain triangle, with a leading-edge sweep of 60°. The advantages of such a wing are structural and mechanical simplicity, ease of achieving low t/c (because of the large chord), low transonic drag-rise and especially low wave drag, good behaviour at AOA beyond that for maximum

The basic design of the McDonnell F-101 Voodoo dated from 1947-48. This example is an F-101B, the two-seat all-weather interceptor version. *PRM Aviation*

lift (with no sudden stall), reduced transonic shift in CP (some might argue about this), and various other gains, such as structural rigidity, making aileron reversal easy to avoid. On the other hand, in 1945 there was no shortage of experts who predicted various dire problems. Most were agreed that a delta would be uncontrollable at high AOA, and that on landing it would either roll off sideways or flip over on its back.

With considerable boldness, the US Army AF placed a contract with Convair in September 1945 for a delta fighter prototype with Mach 1.2 capability. The result was the XF-92A, with 6 per cent 60° wings, and no horizontal tail. First flown on 9 June 1948, it was prudently planned as a subsonic research aircraft (the world's first delta), reaching Mach 0.95. It underpinned the subsequent design of the Convair Model 8-80, built

for the USAF as the F-102. This was a challenging semi-automatic all-weather interceptor, with its busy pilot flying the aircraft and monitoring the radar, autopilot and missile armament. The prime contractor for this weapon system was Hughes Aircraft. Convair had the easier task of producing the vehicle into which everything was fitted. The first of two YF-102s made its first flight on 24 October 1953. This again had 60° 6 per cent wings, but the rather stumpy fuselage housed a massive afterburning

The Convair XF-92A was the world's first delta. Here one wing is tufted to give a qualitative picture of airflow.

These drawings were originally sketched by the author for *Flight* in 1953, to give a simple idea of area rule. The Canberra was not area-ruled, while the P.1, prototype of the Lightning, roughly conformed to it.

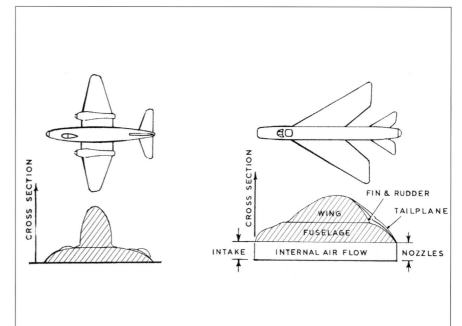

J57, fed by lateral inlets, to leave the nose free for the radar. Severe buffet was met at Mach 0.9, and by 0.93 yaw oscillations were severe enough to endanger the aircraft. With extreme difficulty 1.06 was reached in dives, but Mach 0.99 could not be exceeded on the level.

By sheer chance the NACA had enabled the problem to be solved. First, John Stack had invented the perforated-throat tunnel in which the working section is slotted with longitudinal openings which absorb shockwaves. This at last enabled wind tunnels to explore Mach numbers from 0.8 to 1.2. The previously impossible job of Head of Transonic Tunnels was given to Richard T. Whitcomb, who in 1952 hit upon one of the simplest and most useful aerodynamic rules ever discovered. He found that, for minimum transonic drag, the plot of cross-section areas should form a smooth curve from nose to tail. Thus, if you locally increase the cross-section, for example by adding a wing, you must compensate by subtracting area at the same place, for example by making the fuselage locally slimmer. This was called the area rule, formally stated in December 1953. Some area-ruled aircraft had wasp-waisted or 'Coke-bottled' fuselages superficially similar to those derived at Farnborough by D. Küchemann, but he worked not from area plots but from consideration of the streamlines across the wing. I should add that the 1953 area rule was for the transonic regime. In 1956 Whitcomb published a more complete treatment, extending it to high supersonic Mach numbers where the area plots are considered in sloping planes parallel to the Mach angles, which obviously makes interpreting the rule more difficult.

The area rule arrived at exactly the right time. The presence of the large engine made it impossible to wasp-waist the F-102, so what had to be done was to keep the mid-fuselage as it was and add length and area at the front and rear, in the latter case with prominent bulges on each side projecting aft of the nozzle. With a few other changes, including a cambered (downward curved) leading edge, the production F-102A Delta Dagger eventually matured as an acceptable Mach 1.25 interceptor, but the panic redesign in 117 days has no parallel in aviation. Once having learned how to do it, Convair produced the F-102B, later redesignated as the F-106 Delta Dart. With changes confined mainly to the fuselage, and especially to the inlets and engine, level speed and Mach number were almost doubled.

The XF-103 is discussed in Chapter 8. The Lockheed F-104 Starfighter stemmed from the urgent wish of pilots in the Korean war of 1950–53 to have speed, climb and height to beat the MiG-15, even at the expense of weapons, equipment and much else. The result was 'the missile with a man in it', which in its production form was powered by what can fairly be described as the first properly designed supersonic turbojet installation. The General Electric J79 was fed by lateral inlets and centrebodies (Chapter 7), and exhausted through a con/di nozzle surrounded by a secondary cooling airflow. But the most striking feature was the tiny wing, of only 21ft 11in span and 179sq ft area. It was tapered, but devoid of sweep, had the unprecedented t/c ratio of only 3.36 per cent, and managed to get away with it by having a hinged drooping leading edge and blown flaps. The latter were invented by John D. Attinello at the US Naval Air Test Center. He found that, by bleeding hot high-pressure air from the engine, and blasting it at supersonic speed from thin slits just ahead of the flaps, the airflow could be forced to stay attached, instead of breaking away, even when the flaps were depressed at an angle of 70°. This multiplied the wing lift on landing, though at the expense of running the engine(s) at high power, so in some aircraft something had to be done to cancel the unwanted thrust. Like other aircraft, the Lockheed Model 83, as it was known in the Burbank works, was studied with blowing over the leading edge and tail as well, but designer Clarence L. 'Kelly' Johnson decided against it.

The problems were considerable. Never before had anyone seen a wing with a leading edge that was genuinely sharp, and which had a maximum depth at the root of just 4.2in! Outboard of the blown flaps were the ailerons, where the depth available for the powered control unit was just 1.1in. The answer here was a 'piccolo' actuator with ten hydraulic rams in a single block of duralumin. Some of the worst problems centred on the tail, of T-type, later given an added underfin. Ejecting over this tail looked hazardous, so Lockheed made the seat eject downwards. This led to 15 years of argument, punctuated by a switch to the C-2 upward ejection seat, and finally to Britain's Martin-Baker Mk 7. Early flight testing was hair-raising, but soon confirmed Mach 2.2 capability, with more to come, but the aircraft was red-lined at that point for several good reasons. Armament comprised merely the new M61 Vulcan six-barrel gun and a Sidewinder missile on each wingtip. Despite the flashing performance the F-104A proved to be too limited in capability to have much of a USAF career, but it was followed by a few F-104Cs with three extra weapon or tank pylons. It still could have been a limited programme, but despite not

USAF 53-1787, the Convair YF-102A (Model 8-90), was the first aeroplane in the world to be area ruled. It made all the difference!

First of the Starfighters, the XF-104, had plain inlets feeding a Wright J65 Sapphire turbojet. Supersonic inlets and the J79 engine put the F-104A (illustrated) in the Mach 2 class. *PRM Aviation*

having a home customer Lockheed managed to develop the F-104G version and sell it in huge numbers all over the world, in most cases replacing a British predecessor. The G was structurally reworked, and had new avionics for attack and reconnaissance missions. From it was developed the Italian F-104S all-weather interceptor, which soldiered on until the Eurofighter became available. In 1970 Johnson redesigned the F-104 into the CL-1200 Lancer, with a much bigger high-mounted wing, a low tailplane and a new engine, but it never got off the drawing board.

Republic's F-105 Thunderchief was designed for the USAF as a tactical attack aircraft. It was so big that pilots could walk under its internal bomb bay without stooping, and had to be quite athletic to leap up to inspect the wing-root inlets to the 26,500lb-thrust afterburning J75 engine. These inlets were unique, having a sharp-edged diagonal planform which was the reverse of what seemed normal (Chapter 7). The fuselage was area ruled, the mid-mounted wing had roll-control spoilers, the slab tailplanes were low, and another odd feature was that going into afterburner partly opened the four airbrakes around the nozzle.

Strangely, though Republic Aviation knew a lot about the design of inlets for supersonic flight, they gave the first F-105 plain holes. The production inlet, seen here, was very different. *Philip Jarrett*

The F-106 and 107 have been covered earlier. The North American F-108 Rapier was cancelled before its first flight, but was to have been a very large all-weather interceptor powered by two GE J93 engines burning high-energy 'zip fuel' (Chapter 8) in the afterburners. Many configurations were evaluated, all having the overall shape later to become popular, with a snake-like forward fuselage which at the rear merged into a broad box housing widely separated engines under the very broad wing. This was of delta form, partnered in the final layout by a foreplane and three vertical tails. With a length of almost 85ft, 1,400sq ft of wing, and a cruising speed of Mach 3, the F-108 seemed to indicate that future wars would be fought by radar and missiles at a range of 40 miles at 80,000ft. It didn't work out that way.

Seldom illustrated, the mock-up of the North American F-108 Rapier showed one way to shape a Mach 3 interceptor. Unlike today's F-22, which is similar in size and power, it was not designed for dogfighting, and stealth had not been thought of.

Canada thought along the same lines with the Avro CF-105 Arrow. Avro Aircraft of Toronto had achieved great success with the big CF-100 all-weather interceptor, 692 being built. Despite straight wings and a total non-compliance with the area rule, the CF-100 was dived beyond Mach 1 in 1953. In the same year the RCAF issued a specification for a new interceptor to replace the CF-100 from 1960. The resulting CF-105 first flew on 25 March 1958, powered by twin J75 engines. A tailless delta, it was a monster with a high-mounted wing of 1,225sq ft, a radar with a 40in scanner, and a gigantic internal bay for eleven missiles. The J75-engined prototypes demonstrated a sustained speed of Mach 2.3, and the Arrow Mk 2, powered by the outstanding

Orenda Iroquois engine, would have been even faster. Sadly, the whole programme was thrown on the scrap heap, and the brilliant Avro team with it.

A major factor, which made it easier to cancel the Arrow, in February 1959, was that in April 1957 the British Minister of Defence, Duncan Sandys (pronounced 'sands'), had announced that all manned fighters and bombers were now obsolete, and all the new British programmes were cancelled. Henceforth, he stated, wars would be fought by missiles. Almost the only aircraft that escaped the axe was the English Electric Lightning, which, he said, had 'unfortunately gone too far to cancel'. But the Treasury took care to see that no money was spent on improving the Lightning so that it could win export orders.

I spent the summer of 1949 working as an undergraduate apprentice at English Electric Aviation at Preston. I was excited to hear that, having created the high-flying Canberra bomber, the company was now going to build the only fighter that could shoot it down! At this time the cancellation of the M.52 still warped the outlook of the British government, and it was only with reluctance that public money was spent on playing with subsonic swept

wings and deltas. However, in May 1947 specification ER103 was issued for a purely experimental supersonic aircraft. English Electric responded, and so did Fairey Aviation.

Fairey's Peter Twiss flew the first of two F.D.2 aircraft on 6 October 1954. It was an aircraft of classically simple design, just big enough to house an afterburning Rolls-Royce Avon engine, 322 gallons of fuel and the pilot. The configuration was that of a tailless delta, the thin mid-mounted wing having a t/c ratio of 4 per cent and leading-edge taper of 60°. Instead of elevons, the trailing edge had conventional elevators inboard and ailerons outboard, driven by some of Fairey's first powered-control systems. To give the pilot a view ahead on landing at high AOA, the needle nose was hinged and arranged to pivot downwards through 10° – a so-called 'droop snoot'. Despite having a crude afterburner, with a two-position eyelid nozzle, the F.D.2 proved to have great potential; indeed, Mach 1 was exceeded before the afterburner had been used. In 1955 Twiss, and chief test pilot Gordon Slade, suggested to Sir Richard Fairey that an attempt should be made on the world speed record. The aircraft's owners, the Ministry of Supply, were not merely uninterested; they refused point-

The Avro Canada Arrow was cast in the same mould as the F-108. The missiles were carried in an internal bay bigger than that of a B-29. *Philip Jarrett*

blank to have anything to do with such an attempt, and told Fairey that the company would have to pay for the services of a recording team from Farnborough. Even Rolls-Royce, which had an obvious commercial interest, stated that neither the wing-root inlets nor the engine should be allowed to exceed Mach 1.5. But Fairey battled on, and on 10 March 1956 the first F.D.2 set a record at 1,132mph (Mach 1.73), beating the old F-100C record by 310mph – a margin never equalled before or since. Indeed, the aircraft accelerated firmly throughout each record run, and if it had had more fuel the record could have been set at over Mach 2.

Fairey naturally developed the F.D.2 into a big all-weather fighter to specification F.155T (Operational Requirement 329). This was one of the promising projects cancelled in April 1957 in favour of the non-existent

WG774, the first Fairey Delta 2, was typically British in accomplishing much with little. When it broke the World Speed Record by 310mph, Marcel Dassault took it for granted that it would quickly lead to a fighter. Instead we handed him the world market on a plate. *PRM Aviation*

'missiles'. But, as noted, the other E.R.103 aircraft had 'gone too far to cancel'. The first of two prototypes, known simply as the English Electric P.1, made its first flight on 4 August 1954. It had a strange layout. The fuselage, said by my colleague Rex King to have 'all the shapeliness of a suitcase' had a plain nose inlet feeding an Armstrong Siddeley Sapphire turbojet low down in the mid fuselage, and a second Sapphire in the top of the rear fuselage, the jetpipes (later afterburners) being superimposed. The equally remarkable wing, which passed between the engines, had no less than 60° sweep but hardly any taper, so it was convenient to mount the ailerons transversely so that they joined the leading and trailing edges. The slab tailplanes were mounted low, despite repeated instructions

from the Ministry to use a high T-tail instead. A special research aircraft, the Short SB.5, was built at public expense to prove that there would be trouble with the low tail; to the embarrassment of the Ministry experts, it worked perfectly. The main landing gears were pivoted in the centre of each wing and retracted out and back, the large thin wheels lying near the tips. This cut severely into the limited internal fuel capacity.

Test pilot R. P. 'Bee' Beamont found the P.1 a delight to fly. In May 1948 he had dived an F-86A beyond Mach 1, becoming the first British subject to go supersonic. On the P.1's third flight, on 11 August 1954, he became the first pilot to exceed Mach 1 in level flight in a British aircraft. Subsequently the P.1 was totally transformed into the Lightning all-weather interceptor, with Avon engines fitted with proper afterburners fed by a circular nose inlet with a Ferranti AI.23 radar in the conical centrebody. It pioneered high thrust/weight ratio (often close to unity) and moderate wing loading, which are features of today's fighters. It also had impeccable handling and a rate of climb exceeding 45,000ft/min. The first P.1B prototype Lightning exceeded Mach 1 on its first flight on 4 April 1957, which was the very day on which Mr Sandys – soon

An early English Electric Lightning F1A fighter.
PRM Aviation

to be Lord Sandys – announced the cancellation of all new manned warplanes. The Lightning was reluctantly allowed to continue, but all future additions, such as the electronic auto-attack system and Napier Double Scorpion rocket, were cancelled. Against unrelenting official opposition, English Electric, and from 1960 the British Aircraft Corporation, strove to improve the Lightning, making it a Mach 2 aircraft beloved by its pilots, which was withdrawn from the RAF not in 1962, as predicted in 1957, but in 1989. And we are the folk who pay the politicians!

A contemporary of the Lightning which failed to survive the 1957 'Sandystorm' cancellations was the Saunders-Roe SR.177. This stemmed from the SR.53 which, with the rival Avro 720, resulted from the F.124D specification of 1952. This requirement originally asked for a kind of manned missile, shot off an 80° ramp to climb straight up and fire at enemy bombers, followed by a glide back with empty tanks to land on a skid on the nearest level ground. The officials were outraged when Maurice Brennan, from the Saro flying-boat firm, showed an interest. They were even more furious when he suggested adding a small turbojet so that the fighter could take off and land in the

normal way. After much argument, Saro were awarded a contract for the SR.53 with a Viper turbojet, and a de Havilland Spectre rocket engine burning kerosene in HTP (high-test peroxide). The Avro 720 had a Viper and an Armstrong Siddeley Screamer rocket, burning kerosene in liquid oxygen. Both promised to have fantastic performance at high altitude, the small turbojet merely conferring good range and endurance.

The 720 was cancelled shortly before its first flight, but the first of two SR.53s flew on 16 May 1957, followed by the second seven months later. Painted white, they were attractive aircraft, with a stubby wing with anhedral, leading-edge taper and missile rails on the tips, a finely shaped fuselage with the turbojet above the rocket, and a squat T-tail. As explained in Chapter 7, a turbojet falls away in power at high altitude, whereas a rocket gets more powerful. So attractive did the SR.53 look

The first Saro SR.53 in flight. *PRM Aviation*

that it was developed into the more powerful SR.177, with a Spectre 5A rocket and Gyron Junior afterburning turbojet, arranged the other way round with the turbojet underneath, fed from an efficient multi-shock chin inlet. Calculated performance included Mach 2.5 at anywhere from 30,000 to 70,000ft, an initial climb rate of 55,000ft/min, and the ability to whip into a turn of only five miles radius at 50,000ft and hold it while accelerating from Mach 0.9 to 1.6 in 70 seconds! Today no fighter can get near such capability. The SR.177 was being urgently developed for both the RAF and Royal Navy, with extremely close commonality, when both versions were simply axed in the ridiculous belief that their job of intercepting and identifying intruders could be carried out by some kind of missile.

Though the Saro mixed-power fighters were developed into extremely capable aircraft, operated in the normal

way, they originally stemmed from an Air Staff fixation on a totally useless idea, the rocket interceptor that went straight up, straight down and had no endurance. The same idea interested the French, resulting in the SNCASO Trident. This began in 1948 as the SO.9000, but by the time this flew on 2 March 1953 it had been recast as a research aircraft to support the SO.9050 interceptor. The Trident featured a perfectly streamlined fuselage, its beautiful lines marred only by the canopy and by the fact that at the back the circular section turned into a trefoil (three-leaf clover) shape around the three barrels of the SEPR rocket engine. Amidships was a small, thin, unswept wing carrying a Turbomeca Marboré turbojet on each tip. The latter gave a thrust of only 880lb each, and it says much for Jacques Guignard's nerve that he made the first flight without the rocket. Admittedly, having no rocket propellants cut the weight by about 40 per cent, but the take-off still needed the whole length of the Istres runway. The SO.9050 first flew on 19 July 1955, and had only two rocket barrels but more powerful turbojets, Turbomeca Gabizos of 2,645lb thrust each. Over 100 flights were made at up to Mach 1.8, but basic shortcomings were

short endurance and the ability to carry only one air-to-air missile.

A more successful French fighter builder was Dassault. Starting with the Nene-powered Ouragan (hurricane), this company quickly developed the Mystère, with a thinner and more swept wing. On 28 October 1952 a Mystère II was put into a dive by Capt Davies, USAF, who was evaluating it. To his (and Dassault's) surprise he exceeded Mach 1. Next came the Mystère IV, with more power and a wing that was even thinner and more highly swept, followed via various intermediate types by the SMB2 (Super Mystère B2). Powered by a SNECMA Atar 101G turbojet, rated with afterburner at 9,700lb thrust, this was the first production aircraft in Western Europe (including Britain) to exceed Mach 1 on the level. Then came a sudden change of direction, with the neat Mirage I, a tailless delta first flown on 25 June 1955. Although it had only 3,280lb thrust from two British Viper turbojets, it easily went supersonic in dives.

In collaboration with Fairey, Dassault turned the Mirage I into the Mirage III, first flown on 17 November 1956, powered by a SNECMA Atar. This in turn led to the Mirage IIIA, area-ruled and powered by a more

The Sud-Ouest SO.9000 Trident I after it had been fitted with Viper turbojets on the wingtips and a three-barrel SEPR.481 rocket in the tail. *PRM Aviation*

powerful Atar fed via what the French called 'mousetrap inlets' with half-cone centrebodies. This soon reached Mach 1.66, raised to 1.9 after adding an SEPR rocket under the tail. Dassault was amazed that the British never built a Fairey delta fighter; if you superimpose the F.D.2 on the Mirage III you will see why. To say the Mirage III was a success is an understatement; 1,412 were built of the original tailless delta type, for customers in 20 countries. They have a potential of Mach 2.2 on the level, though working up to this speed (which has to be done on internal fuel) means the fuel state proceeds rapidly towards zero. Today's Mirages feature in Chapter 11.

In 1952 the French Arsenal de 1'Aéronautique, which had produced several excellent piston-engined fighters, was renamed SFECMAS. It received government support

A landing of the second Dassault SMB.1 Super Mystère, the first supersonic fighter in Western Europe.

Despite the name Mystère on the nose, this was actually the very first Mirage, the MD.550 of 1955.

The prototype Mirage III of 1956 still had a long way to go to become the production fighter.

for a planned supersonic delta fighter, the SFECMAS 1501, and, to provide the essential background of data, it built two small research aircraft, the SFECMAS 1402 Gerfaut (gerfalcon). Features included a stumpy hump-backed fuselage, a very thin 58° wing, and small one-piece delta tailplanes pivoted near the top of the large fin. The first, flown on 15 January 1954, was powered only by a 6,173lb SNECMA Atar 101C turbojet, and on 3 August 1954 it became the first aircraft ever to exceed the speed of sound in level flight on the power of a turbojet alone, without afterburner. Two further Gerfauts were flown, reaching over Mach 1.6. In 1955 SFECMAS was merged into Nord, so the next design, the 1500-01 Griffon I, became a Nord aircraft. First flown on 20 September 1955, it was larger than the Gerfaut and had a fixed delta canard and no horizontal tail. The forward fuselage projected from the top of a giant duct housing an Atar plus additional burner rings forming a huge ramjet. It was followed by the Nord 1500-02 Griffon II, first flown on 23 January 1957, which on 13 October 1959 maintained Mach 2.19 in level flight. Largely because of the prior existence of the Mirage III, further work was abandoned.

An even more ambitious series of turbo-ramjet aircraft were designed by René Leduc. All were characterised by fuselages which were nothing more than giant ramjet ducts, the pilot being inside the centrebody of the inlet.

All were air-launched. The first, the O.10, was first flown in October 1947, and eventually reached 500mph on half-power. Via the O.16 and O.21, Leduc moved on to the O.22 fighter, planned to reach Mach 2. This made 33 flights on turbojet power alone, but in 1957 government funding was withdrawn, and no Leduc aircraft ever exceeded Mach 1.

A unique tailless delta was created by the independent Swedes, and in my view it was one of the best fighters of the 1950s. The Saab J35 Draken (dragon, or dragon-kite) was preceded by the Saab 210 research aircraft, which was built to prove the radical shape, called a double-delta. This in turn stemmed from the bold decision to arrange everything axially from front to rear, whereas in previous tailless aircraft the parts had been disposed transversely. When the first J35 flew, on 25 October 1955, it looked rather like a paper dart, with a leading-edge sweep of 57° outboard, and the amazing angle of 80° inboard. Production Drakens had almost the same engine as the Lightning, a Swedish-made afterburning Avon, but the Saab aircraft had half as many engines and much

Powered by an Armstrong Siddeley Adder, the little Saab 210 research aircraft explored the aerodynamics of the projected Draken. It first flew in December 1951. The nose and far wing are tufted.

more internal fuel. The Draken sustained an excellent programme, leading to advanced missile-armed versions with outstanding radar and infra-red sensors. Maximum speed varied between 1.8 and 2 (1,320mph), and it is interesting that the 606th and last Draken, delivered in 1975, had a maximum weight almost precisely double that of the first.

In the mid-1950s the Soviet design bureau of A. N. Tupolev worked on several supersonic aircraft, including the Tu-98 and Tu-105 (Tu-22) bombers, described in Chapter 8. A closely related aircraft was the Tu-102,

with service designation Tu-28. The version that went into production was the Tu-28P interceptor (Tupolev OKB number, Tu-128), and on most counts this was the largest fighter ever to be put into service anywhere. Over 89ft long, and weighing 100,000lb when loaded with maximum internal fuel and four large missiles, it could reach Mach 1.65. Its size reflected the fuel capacity needed to defend the world's largest country.

In parallel, the lessons of the Korean War resulted in the careful testing of two new fighter configurations in 1955–6. Both had swept slab tailplanes, but one had a slender wing swept at 62° on the leading edge, while the other was a pure delta with a leading edge angle of 57°. MiG and Su prototypes were built in both configurations, in different sizes. The aircraft built in by far the greatest numbers, over the longest period, was the MiG-21. Like the contemporary F-104, this began as the simplest and most basic fighter imaginable, armed only with two small missiles. Later a gun and bomb/rocket pylons were added, and improved radar, but an Indian pilot still called the MiG-21PF 'my supersonic sports car'. Despite this, well over 10,000 were built in the USSR, with production of Chinese derivatives still proceeding, and more countries use MiG-21s (at least 45) than any other supersonic aircraft.

The same configuration was used for the Su-9 and Su-11 all-weather interceptors, matched to engines of roughly twice the thrust. Amazingly, the same configuration was then used for the Su-15 interceptor, powered by two of the same Tumanskii engines as the MiG-21. The highly swept conventional wing was used for the Su-7 day fighter, originally intended 'to shoot down the F-100', which actually matured as the Su-7B ground-attack aircraft. Though brutishly simple, and almost unbreakable, the Su-7B had beautiful handling and was very popular. Its swing-wing derivatives appear in Chapter 10.

Throughout the 1960s the name Foxbat sent shivers down many spines in the Pentagon. This was the NATO reporting name for the MiG-25, which began its career by capturing various speed records in April 1965. Many further records followed, and to this day one MiG-25 mark – a zoom to 123,523ft – remains unbroken. Yet in fact the MiG-25P, the interceptor version, has always been an aircraft of doubtful cost effectiveness.

It was designed to shoot down the B-70 (Chapter 8), and when that bomber was cancelled the fighter was continued. The basic MiG-25 has a high wing with anhedral and sharp taper on the leading edge, and anti-flutter tubes attached to the broad tips. The 4.4 per cent wing is remarkably simple, with a fixed leading edge, conventional ailerons near the mid-point of the trailing edge, and simple plain flaps inboard. Bearing in mind the wing loading of 135lb/sq ft, and the poor lift of such a thin wing at low speeds, it is clear that nobody was bothered about needing take-off and landing distances of around 2km (6,600ft) and a take-off speed of over 200mph Moreover, with such a high wing loading, any kind of dogfight agility was out of the question, and at the maximum level speed of Mach 3.2 (2,115mph) the

The Tupolev Tu-128 (28P) was one of the largest fighters of all time, with a length of over 89ft and take-off weight of nearly 100,000lb. *PRM Aviation*

aircraft was obviously going to proceed in a straight line. With its four giant missiles, carried externally, the speed in service is limited to Mach 2.87, but this is still much faster than any other fighter, even without missiles. Thus, the original design mission was to kill hostile aircraft from a distance.

The engines are two Tumanskii R-31 turbojets, simple engines each rated at 27,010lb with maximum afterburner. They are fed from huge 2-D inlets of the type pioneered by the Vigilante, with variable upper walls and a hinged lower lip. At high Mach numbers water-methanol is injected into the inlets via pipes running externally forwards from the leading edge of the wing. A total of 3,830gal of special non-coking fuel is housed in structural tanks in the fuselage, in saddle tanks around the ducts, and in unusual integral tanks in the wings, forming the inner volume inside double skins. Only the outer skin carries flight loads. When the MiG-25 was first seen in 1967 its twin fins, canted outwards, attracted much attention, though today such tails are common. They were adopted partly because, together with the quite large ventral fins, they are structurally compatible with the twin engines, partly because they can be aeroelastically preferable to large and flexible single fins, and not least because at all times at high AOA they give superior positive directional stability, and in yawed flight there is always one vertical tail operating in fairly clean air. At high supersonic Mach numbers each fin shock front passes behind the adjacent fin, eliminating mutual interference. There are other complicating factors but, as the Vigilante was modified to have a single fin and the F-108 was never built, the MiG-25 was the first supersonic aircraft after the Lockheed A-12 to appear with twin fins.

Almost all of the MiG-25 airframe was arc-welded nickel steel. The Mikoyan bureau had since 1945 favoured welded joints, and the MiG-25 represented a peak of such construction, the new material being adopted because of its retention of strength when soaking at temperatures higher than 300°C. Leading edges, however, were of machined titanium, while certain cooler items, including the ailerons and rudder, were of light alloy. The standard armament of the MiG-25P interceptor version comprised four large air-to-air missiles on under-wing pylons. In conformity with Soviet practice these were normally split 50/50 between the R-40P version with semi-active homing, the target being illuminated by continuous-wave (CW) transmitters at the front of the MiG-25's wing-tip

pods, or the R-40T with passive infra-red homing. No gun was fitted, the MiG-25 being useless in close combat. On the other hand, the basic Ye-266 design was ideal for a tactical reconnaissance aircraft, and several MiG-25R versions were developed, some with a sideways-looking airborne radar (SLAR) and five cameras, and others with a larger SLAR and no cameras. Infra-red linescan has also been carried. Oddly, the reconnaissance version has a modified wing, with a span of 44ft instead of 45ft 9in, and a straight leading edge (the fighter has a slight kink). Even stranger, the smaller wing is said to give a higher ceiling of 88,580ft instead of 80,000, and even more curious is that the reconnaissance version has a shorter combat radius, 560 miles instead of 700.

In 1976 a MiG-25 pilot defected to Japan. He revealed that an improved fighter was being developed, and in 1981–8 about 200 MiG-31s were delivered. These retain portions of the MiG-25 fuselage and tail, but are totally different aircraft. Redesigned inlets feed the impressive D30F-6 engines by Reshetnikov, each of 41,843lb. The wings are completely new, with greater chord and sweep and much larger flaps and ailerons. The main landing gears are redesigned, with twin wheels on each side, the front wheels being inboard of the legs and the rear wheels outboard, each tyre being much smaller than on the MiG-25. Radar and systems are completely new, and a weapon-system officer is now carried in a second cockpit behind the pilot. Combat persistence has been doubled by adding four pallets under the fuselage for the large AA-9 missiles, leaving the two or four wing pylons free for further AA-9s or for up to 12 small AA-8 dogfight weapons. A six-barrel 23 mm gun with 260 rounds is installed in the right side of the rear fuselage. The peak speed has been lowered to Mach 2.55 (1,685mph), still ample to catch almost any known target. The structure and wing aerodynamics have been upgraded to give the new interceptor the ability to fly at full throttle and pull tight turns all the way down to sea level. Combat radius is said to have been increased to 1,305 miles, despite the high fuel burn of what is still the fastest fighter currently available.

The only aircraft with equal speed is another interceptor of the PVO, the Soviet air-defence organisation. The Su-15 began life as a twin-engined version of the Su-11 in 1962, when the menace of the B-70 had not quite evaporated. The requirement was for an interceptor able to hold Mach 2.5 and home with automatic guidance towards its target. Vladimir Ilyushin flew the first T-5 prototype in about 1963, the aircraft being characterised by the small size of the delta wing in relation to the large fuselage. Considerable development took place, one aircraft (Su-15VD) having three lift engines amidships as in contemporary MiG testbeds, though no example was flown with VG wings. Eventually the main production version was given an improved wing, with span extended from 27 to 30ft by reducing sweep on the outer panels, and thus, by fitting into the 70ft fuselage two Tumanskii R-13F2 turbojets each with an afterburning thrust of 14,800lb, achieved the remarkable thrust/weight ratio of 1.25. The lateral inlets were fully variable and claimed to achieve close to 100 per cent efficiency in every regime of flight. Twin fins were studied, but found unnecessary. Though basically a standoff killer, the Su-15 also has close-combat capability and can carry twin GSh-23L gun pods.

In summer 1990 the 329th and last Saab 37 Viggen was handed over to the Swedish air force. System 37 began in 1962, the requirement being for a supersonic combat aircraft produced in several closely related versions for different missions, all with STOL capability and able to fly missions electronically linked with the national command and control system. Design features included an aft-mounted delta wing with a fixed leading edge and trailing-edge elevons, a delta foreplane with trailing-edge flaps, a single large augmented turbofan engine with reverser, plain fixed lateral inlets with ducts passing above the wing, and inward-retracting main gears with two small wheels in tandem stressed to accept no-flare landings at a sink rate of 16.4ft/sec on to unpaved strips. The first and most numerous version was the AJ37 attack aircraft. Then came the SF37 and SH37 for reconnaissance and the tandem-seat Sk37 dual trainer. Finally, after much further development, came the JA37 interceptor, with the same extended fin with swept tip as on the trainer (all Viggen fins can fold over to the left) and four elevon power units instead of three. Despite the simple inlets all Viggens can reach Mach 2, the fighter achieving over 2.15. The foreplane generates huge vortices across the wing, and using the no-flare technique Viggens can take off and land using a 1,640ft strip, microwave ILS being put down in minutes to give precision landing guidance. This is survivable airpower, unlike NATO's.

A contemporary of the Viggen was the Anglo-French SEPECAT Jaguar, designed for Mach 1.6. This began life as a low-powered trainer and then, thanks to British pressure, emerged as a very capable attack aircraft which, like the Viggen, has plain inlets and can operate from short

unpaved strips. Another aircraft with fixed inlets which began as a trainer and then grew up is the F-5. Northrop's unbuilt Fang fighter of 1955–7 led to the T-38 Talon, built in numbers for the USAF and still the only supersonic undergraduate pilot trainer in the world. Powered by two afterburning GE J85 engines, this led in 1959 to the F-5 Freedom Fighter, notable for going into production despite having no order at that time from its own country. This in turn was developed into the F-5E Tiger II, which is still a useful type in many air forces. Altogether 3,806 aircraft in the T-38/F-5 family were sold, to 26 countries. Northrop was sad that what could have been a further generation, originally designated F-5G and later F-20A Tigershark, failed to find a customer. This had a further improved airframe and the vastly greater power of a GE F404 engine, as fitted to the F/A-18 (described later), but with one engine instead of two. It also had a modern 'glass cockpit' with multifunction colour displays instead of dial instruments (indeed, McDonnell Douglas accused Northrop of stealing the technology from the F/A-18). The Tigershark had combat readiness, flight performance and agility, which in some respects exceeded that of any contemporary Western fighter.

In passing, I must record that in 1959 a Martin P6M-2 SeaMaster, an enormous four-engined flying boat, was dived to over Mach 1.05. Other large aircraft dived beyond Mach 1 at around the same time included a Handley Page Victor bomber and a Douglas DC-8-40 jetliner. None of these could really be described as supersonic aircraft.

Making supersonic fighters compatible with aircraft carriers was obviously difficult. Lippisch ideas were used by Ed Heinemann, the great leader of Douglas's Navy plant at El Segundo, in designing the tailless F4D (later F-6A) Skyray. The wing was a swept surface of low aspect ratio, blended into the short fuselage, the thick roots housing most of the 533 gallons of fuel. At the aft end of these roots, inboard of the elevons, were giant pointed trimmers, locked in the up-position for take-off or catapult launch. On the leading edge were automatic slats, and I recall two top British designers marvelling that such a thing was possible on so thin a wing. Even powered by a J57, the Skyray never got beyond Mach 0.98 on the level, but it was developed into the F5D Skylancer of 1956, which, merely by reducing the t/c ratio, sharpening the canopy and inlets, and lengthening the fuselage, reached Mach 1.44. The NACA calculated that double-shock inlets would give the Skylancer Mach 1.8.

The XF5D-1 Skylancers were highly supersonic developments of the Skyray, but this was one competition Ed Heinemann didn't win.

Grumman's F11F-1 Tiger was an attractive little fighter, which exceeded Mach 1 on the level without using afterburner.

The Skyray was designed before the area rule had been discovered. The first Navy fighter area-ruled from the start was the Grumman Tiger, originally designated F9F-9 but then more correctly as the F11F-1. Small and neat, the initial version, powered by the Wright J65 (licensed Sapphire), was the first American aircraft to exceed Mach 1 without using afterburner. Later, two were modified with J79 engines and reached Mach 2 on the level. Grumman delivered 199 of the original model, and on 21 September 1956 one managed to shoot itself down. Diving at high speed, it eventually overtook the shells from its own 20mm cannon, and these caused severe engine damage, resulting in a belly landing.

By starting later, Chance Vought did even better. The requirement was issued in September 1952, and the XF8U-1 Crusader prototype flew on 25 March 1955, easily exceeding Mach 1 on the first flight. The design incorporated area ruling from the start, so, instead of having a wasp waist, the fuselage had the same 'suitcase' look as Britain's P.1. The afterburning J57 was fed by a sharp chin inlet, the conical radar above giving a two-shock configuration (Chapter 7). A unique feature was that the wing was mounted on top of the fuselage on hinges, so that, for take-off and landing, it could be

set to +7°. This kept the fuselage horizontal, giving a good pilot view ahead, and enabling the main gears to be short, folding into the fuselage. The F8U, later redesignated F-8, was a brilliant design which sustained a production run of 1,262, ending with 42 for France with more powerful high-lift systems for operation from small carriers. These remained in service until 2008 (see Rafale, Chapter 11). On 2 June 1958 Vought flew the first XF8U-3, with a thinner wing, huge folding ventral tail fins and a J75 fed by an efficient variable chin inlet. Speed jumped to 1,580mph (Mach 2.32).

Reflecting the USAF view of the F-107A, the Navy called the XF8U-3 'by far the best airplane we ever canceled'. It was cancelled because it had the misfortune to run up against something exceptional: the McDonnell XF4H-1, later redesignated F-4A, better known as the Phantom II. By designing it by themselves, the St Louis team created a masterpiece which was bought not only by

Vought's impressive XF8U-3 Crusader III, with ventral fins folded and the wing in the high-incidence position. Beyond is the second XF8U-1.

The very first Phantom II, then called the McDonnell XF4H-1. *Philip Jarrett*

Visible condensation in the supersonic expansions around an F-4J of US Navy test squadron VX-4 at Pt Mugu. The Mach number would be about 0.95.

the US Navy but also, unprecedentedly, by the Air Force. This is despite a jagged 'upside-down' appearance which belied the tremendous performance conferred by twin GE J79 engines fed by well-designed variable side inlets, and with the variable afterburner nozzles projecting aft of the fuselage structure. The wing had sharp taper on the leading edge, a t/c ratio of 6.4 per cent at the root, thinning off to 3 at the tip, and blown drooping leading edges (later replaced by slats) and blown plain flaps. The folding outer panels had a startling 12° dihedral, with their leading edges extended forward to form large discontinuities called a dogtooth, or a sawtooth. At high AOA these forward projections generated a powerful vortex, which writhed back across the wing to keep the flow energetic and prevent breakaway. Similar discontinuities are seen on many wings, and also on tailplanes (notably that of the F-15). The F-4 tailplanes had no dogtooth, but the unusual anhedral of over 23°. Despite these odd features the F-4 became the standard against which other contemporary fighters were judged, and 5,211 were built.

During the Second World War many experimental seaplane fighters were built, and in Japan several types were put into service. Experience showed that such aircraft were unlikely to win in combat, and this view was

The XF2Y-1 Sea Dart is seen here with one of the twin-ski arrangements, in this case toed sharply inward.

reinforced by the British Saro SR.A/1 twin-jet flying-boat (US = boat seaplane) fighter of 1947. Yet the US Navy, perhaps rightly, wished to see what could be done with a flying-boat fighter designed for supersonic speed. At rest it would float in the water. Accelerating under the thrust of afterburning turbojets, it would soon rise out of the water, and ride on one or more hydroskis. After takeoff the ski(s) would be retracted. The result was a series of Convair F2Y Sea Dart fighters which came remarkably near to success. The first, the XF2Y-1, was underpowered with twin Westinghouse J34 engines, but began its flight test programme on 9 April 1953. Four YF2Y-Is were powered by the 6,000lb thrust Westinghouse J46, and on 3 August Charles E. Richburg dived one beyond Mach 1. What killed the project was a combination of protracted ski problems, a highly publicised mid-air break-up, and a general loss of interest in the concept. But several designers have looked at the hydroski fighter since 1975, partly because it is difficult to incapacitate their airfield.

Chapter 6

FASTER AND FASTER

One of the many paradoxes of aviation is the fact that, whereas in the 1950s designers were striving to reach Mach 2, 3 and above, and even combat aircraft were being designed to reach Mach 3, today there are no high-Mach research aircraft (except for the US NASP, Chapter 12), and most air-combat training is done at about 400 kt. Yet another paradox is that the fastest thing in today's sky is the Shuttle orbiter (Chapter 12), a huge bluff-fronted vehicle which appears to be totally unstreamlined.

Most of the responsibility for starting the push up the Mach scale with manned aircraft rested with the improved versions of the USAF Bell X-1 and the Navy Douglas Skyrocket. Before turning to these it is worth having a brief look at two missile programmes. They were influenced by the pioneer cruise missile, the German Fi 103 or 'V-1'. Both were programmes of the USAF, but while the Bomarc was intended to swoop down and kill enemy aircraft, the Navaho was planned to carry a massive thermonuclear warhead to distant cities. Both tasks were considered to require an air-breathing vehicle capable of sustained highly supersonic speed.

As briefly noted later in this chapter, the German A.4 rocket opened the way to intercontinental ballistic missiles and ultimately to spaceflight, but in the immediate postwar period there was no certainty that any of this could be accomplished. Accordingly the USAF launched various winged cruise-missile programmes, of which the most ambitious was to become SM-64A Navaho. This was to be a huge vehicle, over 95ft long and weighing 292,000lb at launch. It was to be fired almost vertically by a boost stage powered by three rocket engines far more powerful than anything previously attempted. These engines, burning lox and RP-1 (kerosene), were developed by Rocketdyne, a division of North American Aviation (NAA) specially created for the task in 1955, and which was to enjoy a US monopoly of rocket engines in the very largest sizes up to the present day. The booster's thrust of 415,000lb launched the mated vehicles off the pad, and accelerated the Navaho into supersonic flight at high altitude. The missile's own Curtiss-Wright RJ47 ramjets (Chapter 7) were then lit, to propel the missile at 1,980mph (Mach 3) over a range of up to 6,325 miles.

This was the first vehicle planned to cruise over long distances at supersonic speed. Clearly the number of configurations and propulsion systems was open-ended, and NAA, which was also prime contractor on the Navaho, built a series of X-10 pilotless vehicles – recoverable aeroplanes – to prove the propulsion, flight control and systems. All had the configuration finally adopted for the missile: a large needle-nosed fuselage riding on a low-mounted 60° delta wing of only 2 per cent thickness over the outer portions, with a canard foreplane of the same shape. Side inlets fed engines mounted side by side above the wing. On the sides of the engine bays, in other words on the rear fuselage, were two tails canted out at 24°. This was the first time the world had seen such tails, which are today common on fighters. They were used for yaw control, all main pitch/roll control being by elevons on the wings and the canards being used for major pitch demands (e.g. to raise the nose steeply on take-off, the elevons being neutral), and for pitch trim at all times. Another feature that was to become common was that all the horizontal surfaces were cut off at the cruise Mach angle of 2.05 to avoid the tips having to lie outboard of the tip shock in a separated-flow region. The X-10 was powered by two Westinghouse J40 afterburning turbojets; thirteen were built and did a lot of useful flying at Edwards and Cape Canaveral, under remote radio control. The much bigger Navaho missile also made 11 flights before cancellation

Though unmanned, like the X-10, the Bomarc missile pioneered several technologies, including pulse-doppler radar and ramjet cruising at Mach numbers nudging 4.

Two of the stretched and improved X-1 family, each proclaiming its identity. When this photo was taken, probably in August 1954, the X-1A bore fifteen mission symbols.

in 1958. The chief reason for cancellation was the development of the even more 'unstoppable' ICBM, as outlined later.

The Bomarc got its odd name from Boeing, the prime contractor, and Michigan Aeronautical Research Center. Boeing began SAM (surface-to-air missile) studies in 1945 with the GAPA (ground-to-air pilotless aircraft). As finally flown in 1955 the IM-99 Bomarc had a 44ft untapered fuselage with a high wing and conventional tail, all surfaces having broad tips cut off at the angle corresponding to the cruise Mach number of 3. It was launched vertically by a rocket which accelerated the 16,000lb vehicle until the two underslung Marquardt RJ43 ramjets could ignite and take over propulsion. The complete wingtips were pivoted to form ailerons, the horizontal tails were slab elevators, and the complete top of the fin was pivoted to act as the rudder.

To support the Bomarc programme Lockheed was contracted to build a supersonic unmanned test vehicle, the X-7. Eventually this was assigned many tasks, and including Navy XQ-5 target drones no fewer than 61 vehicles of this family were produced, performing at least 100 missions. The X-7 was characterised by a needle-nosed tubular fuselage with a stubby mid-mounted wing. In the X-7A-1 this had modest taper and a broad tip, as in the same maker's F-104; in the X-7A-3 and XQ-5 the wing had greater root chord and extremely sharp taper, as in today's ATFs. There were small ailerons at the tips, and the horizontal tail comprised slab elevators pivoted to the fin (there was no rudder). Launch, from the ground or from a carrier aircraft, was boosted by a tandem rocket or two underwing rockets, until the large underslung ramjet engine could be started. Various ramjets were fitted, including scaled-down versions of the Navaho's RJ47, but by far the most common were versions of the Bomarc's Marquardt RJ43, of 28in diameter. Cruising speed varied from about Mach 3 up to just beyond Mach 4. Though mainly intended to prove the engines, the X-7s were also used for much research into other problems at up to Mach 4, not least

being how to design airframes to penetrate what by 1955 was being called 'the heat barrier'. X-7s were the first vehicles able to cruise at high Mach numbers for long enough for the structure to 'soak' – in other words to heat up to its ultimate stagnation temperature. Leading edges and other parts, to say nothing of the ramjet, could reach temperatures of up to 420°C (800°F). The ruling airframe material of the X-7 was Type 4130 heat-

treated steel, and in most missions the outer surface was blackened by the temperature except where internal structure could conduct heat away. Without publicity the X-7s made many later programmes much easier. For the record, the longest flight was 552 seconds, and the peak Mach number 4.31, or 2,881mph.

As noted in Chapter 4, once the Bell X-1 had flown faster than sound it was a relatively simple task to find ways to improve it. The resulting X-1 A, X1B and X-1D were perhaps most remarkable in that the propulsion system, wings and tail were left unchanged. The availability of the turbopump feed clearly removed the previous limitation on propellant capacity, but the bomb bay of the B-29 or B-50 carrier aircraft limited the possible increase in fuselage length to 4ft 9in. This was still sufficient to increase tankage to 475 gal of alcohol and

The X-1A drops from its carrier aircraft, the same B-29 as used for the first XS-1 flights. The rocket engine is about to ignite.

416gal of lox; it also added a parallel section amidships which, in the author's view, improved the appearance. This was further enhanced by fitting a normal fighter-type windscreen and canopy, the latter swinging up and to the rear on parallel links. Whereas the original X-1s had had a control column with a two-handed rotary aileron yoke on top, the second-generation aircraft all had normal fighter-type sticks. Shortly after the start of flying an ejection seat was installed, but there were few other changes.

The first to be completed was the X-1D, but this had a brief career. Jean 'Skip' Ziegler made a gliding flight on 24 July 1951. A month later, on 22 August, it was about to be launched from a B-50 on its first powered flight when a severe explosion occurred. Pilot Lt Col

'Pete' Everest quickly left the cockpit, and the X-1D was jettisoned to destruction. Ulmer leather was the cause. Next came the X-l A, making its first glide flight on 14 February 1953. Ziegler took it up to Mach 0.93 in April, before handing over to Yeager, who reached 1.15 in November, followed by successive flights in December 1953 to 1.5, 1.9 and finally 2.44, or 1,650mph The last flight overtook the previous record of the Navy's Skyrocket (see later), but it almost ended in disaster. At maximum speed at about 70,000ft Yeager suffered inertial coupling, exactly like that which afflicted the F-100 (Chapter 5). The X-1A tumbled and gyrated so violently that Yeager became unconscious, eventually coming to and painfully working out that the X-1A was in an inverted spin. On 8 August 1955 this aircraft became yet another victim of the infamous ulmer leather, exploding in flight. This left only the X-1B, first flown on 24 September 1954. On 2 November 1954 Col Everest reached Mach 2.3 in this aircraft; then it was turned over to the NACA, who fitted an upward-hinged canopy and, later, slightly extended wingtips housing the hydrogen peroxide thrusters of an exo-atmospheric control system. The last four missions, ending in January 1958, were flown by the then-unknown Neil Armstrong.

One more X-1 was built, though it failed to demonstrate its calculated potential. The X-1E was in fact a complete rebuild of 46-063, the second X-1. Its main difference was that Stanley Aviation, the Denver-based company formed by Bob Stanley, former Bell Director of Engineering, produced wings and tail surfaces of only 4 per cent thickness, the lowest then achieved with a manned aircraft. Other changes were that the vee-type windscreen and LR8 four-barrel rocket were similar to those of the Navy Skyrocket. The X-1E made 26 flights in 1955–8, and reached Mach 2.24, and today is 'gate guardian' at NASA Dryden Flight Research Center.

Having said so much about the Skyrocket, I had better bring it into the book. In my view it was perhaps the most beautiful aeroplane ever built. Like the X-1 it had its genesis in the big gatherings at NACA Langley which began on 15 March 1944. Competition between the US Army and Navy had been more common than collaboration, and in March 1945 Douglas Aircraft's Navy plant, the El Segundo Division, began talking to the Navy about high-speed research aircraft. Chief Engineer Ed Heinemann continued talking, while some of his men combed Germany for transonic data. In June contracts were signed for three aircraft to reach as near Mach 1 as possible, and three more – a bold decision – aimed at Mach 2. The NACA participated, and agreed to collect all data.

The next number in Douglas's project list was 558, and the two types of aircraft became the D-558-1 Skystreak and D-558-II Skyrocket. They had little in common. The D-558-1 was basically a conventional aircraft with a tube-like fuselage housing a General Electric TG-180 (J35) turbojet of 4,600lb thrust. Later this engine became the Allison J35, of 5,000lb, but no afterburner was fitted. The tiny straight wing was mounted low, and sealed inside to form an integral tank, there being little room for fuel in the stovepipe fuselage. The entire nose was made jettisonable, complete with the front part of the extremely small rounded canopy. Painted bright red, so that someone called it the Crimson Test-tube, the first Skystreak began flying on 28 May 1947. It generated masses of data, besides setting world speed records in August 1947 at 640.7 and then 650.8mph, but it never got beyond Mach 0.99.

The sleek D-558-II was a different kind of animal. Heinemann kept Bob Donovan as project engineer but assigned Kermit Van Every as aerodynamicist. Van Every had masterminded the odd shape of the Skyray Navy fighter. Unlike the Bell X-1, it was decided to design the Skyrocket primarily for conventional take-off. The span of 25ft was the same as for the D-558-1, but the wing was totally different. It had 5 per cent thickness and a normal profile with a rounded leading edge, 35° sweep and anhedral. In a design tour de force automatic slats were

The first D-558-II Skyrocket, 37973, is seen taking off from the ground under the power of boost rockets. *PRM Aviation*

The Skyrocket was a singularly beautiful aircraft, even in its original form with flush jet inlets under each side of the forward fuselage.

fitted, an amazing achievement on so thin a wing. The tail was conventional but swept at 40°, with the powered tail-plane halfway up the fin. The fuselage had to be big to house the 3,000lb Westinghouse J34 turbojet and its 216 gal of fuel (which was not kerosene but petrol), the Reaction Motors XLR8-RM-5 rocket (very similar to the X-1 engine) in the extreme tail, the 142gal of lox in a lagged sphere ahead of the wing, 160gal of alcohol aft of the wing and the three units of the landing gear. The J34 was fed by flush inlets low on each side of the forward fuselage and

exhausted through a nozzle under the rear fuselage well ahead of the rocket. As in the Skystreak the nose section, with a sharp vee windscreen, could be jettisoned.

John F. Martin made the first flight from Muroc on 4 February 1948. No rocket was installed, and he naturally wished for more power. A year later a workable XLR8 was received and Gene May began explorations beyond Mach 1, soon joined by Bill Bridgeman. By the summer of 1951 air launches were being made from a Boeing PB2B-1S (B-29). In August 1,238mph was reached, followed a few days later by 79,494ft. These figures were set by a Skyrocket with the J34 removed and the rocket tankage increased to 287gal lox and 315gal alcohol. Work continued, with 83,235ft reached in August 1953 and Mach 2.05 (about 1,360mph) on 20 November 1953, the first time Mach 2 had been exceeded. Bridgeman did both.

The Navy called the Skyrocket carrier a PB2B. Here it has just let go of the second D-558-II, 37974, with the turbojet removed and much more rocket propellant. *PRM Aviation*

USAF No 46-680 was the first of the two Republic XF-91 Thunderceptors. They were unconventional in almost every feature.

A familiar artist's impression, but still the best to show almost the final configuration of the amazing Republic XF-103.

With strong support from a small number of people in the Navy and NACA Heinemann's team next planned the D-558-III. To describe this as merely the next stage is an understatement; indeed it could have been the first aerospace plane, and so is discussed in the final chapter.

We have seen repeatedly how formidable was the 'sound barrier' when designers were uncertain how to tackle it, and how it melted away with increasing knowledge. Thus, by 1953 even the designers of fighters were looking towards Mach 2. They knew how to go 'through the barrier', and beyond this – discounting such unexpected problems as inertia coupling – it

was simply a matter of designing efficient multishock inlets and keeping thrust ahead of drag. In general, things went beyond prediction, partly because increasing speed resulted in improved inlet pressure recovery and thus greater thrust. For example, the F-104, F-106, F-107 and XF8U-3 all easily exceeded Mach 2, and had to be restrained because of their aluminium-alloy structures.

One company aimed even higher. Republic Aviation, on Long Island, had been pipped to the post by NAA's swept-wing F-86, but then embarked on two of the most remarkable fighters of all history. The first was the XF-91 Thunderceptor, first flown on 9 May 1949. Planned as a target-defence interceptor, it ended up with quite a respectable range (1,171 miles) despite having both turbojet and rocket propulsion. It was unconventional in almost every way. Most odd of all was the wing, which not only had inverse taper and thickness but was pivoted to the fuselage so that its incidence could be varied. This was despite the fact that the main landing gears, each with two wheels in tandem, were attached to the wing and in fact retracted outwards to be stowed in the huge tips. Later an XLR11 rocket was installed, as in the X-1, but with the four barrels arranged in a vertical row, two above the J47 afterburner and two below. Mach 1 was first exceeded in December 1952, but by this time the XF-91 had long since been overtaken by later developments.

One of these developments was an amazingly advanced project for an all-weather interceptor to meet USAF requirement MX-1554. This was won by the Convair F-102, as already noted, but Republic's proposals were so bold as to lead to a next-generation aircraft to replace the F-102. Designated XF-103 by the USAF, Advanced Project 59 was to lead to the fastest and most formidable stand-off interceptor imaginable. At first there were three choices of propulsion system: turbojet plus rocket, afterburning turbojet, and turbo-ramjet. During 1951 the third scheme was chosen, and the XF-103 went ahead, for service in 1959–60. This was a very long timescale for those days, which says a lot!

Over the next six years the basic design changed little, but in detail the XF-103 was repeatedly altered. The one thing that stayed broadly unchanged was the propulsion system. A giant inlet under the fuselage (fitted with retractable screens to catch anything thrown up by the nosewheels) fed air at take-off to a Wright J67 turbojet which, via a short upward-sloping connection, discharged into an afterburner. The J67 was an Americanised Bristol Olympus, rated with afterburner at 22,100lb. This was to take the XF-103 to supersonic speed at 35,000ft, where over a period of seven seconds the inlet would close off the J67 and pass all air direct to the afterburner, converting it into a Wright XRJ55 ramjet rated at up to 37,400lb at about Mach 2.6 at 35,000ft. The inlet and nozzle both had to be fully variable in profile and area, and so initially both were designed to be square in section. Later the nozzle was made circular.

The aircraft itself was made of titanium and steel, and had an enormous fuselage which by 1953 had grown to a length of almost 82ft. Halfway along was the relatively thin mid-mounted delta wing, the tips of which formed huge horn-balanced ailerons. The tail was also delta-shaped, comprising a fin and huge horn-balanced rudder and two powered tailplanes. Under the tail was a fold-down ventral fin. The wing leading edge was straight and fixed, and inboard of the ailerons were double-slotted flaps. The t/c was 3 per cent, which posed severe

challenges. The monster fuselage housed 1,900gal of JP-4 fuel, with provision for drop tanks. At the front the pilot looked out through a periscope, while behind him were internal bays for the missile armament. Originally the latter included 36 Mighty Mouse rockets, but these were later replaced by six Advanced Falcon guided missiles.

Sadly the XF-103 programme was cancelled on 21 August 1957. This was at the height of the debate over whether, as Britain proclaimed, missiles had already made all fighters obsolete. In fact there were also other reasons. In its final form the XF-103 would have weighed 55,780lb and been able to cruise at 1,980mph (Mach 3) at 70,000ft. At full ramjet power the limit was predicted to be Mach 3.63 at 75,000ft and 3.7 (2,446mph) at the ceiling of 80,000ft. So far as I know no fighter has been designed to fly so fast since that time, and in any case Mach 3.7 means you fly in a straight line.

In the early 1950s both the USA and Soviet Union planned uninterceptable reconnaissance aircraft. The Lockheed U-2 relied on height only, but the Tsybin RSR was designed from 1955 to reach 3,000km/h (1,864mph, Mach 2.8) at 30km (98,400ft). P. V. Tsybin wisely underpinned this challenging aircraft with a research aircraft, the NM-1 (scientific model 1). It had a conventional tailed configuration, with a fuselage over 87ft long riding on stubby sharply tapered wings carrying a Mikulin AM-5 turbojet on each tip. Problems were encountered and the NM-1 did not exceed 304mph on test.

Another research aircraft, on which far more money was spent, was the Bristol 188. This was designed to meet Experimental Requirement 134, issued in 1954, calling for a conventional aircraft able to soak at Mach 2. The need to take-off and land normally, and hold Mach 2 for long enough for airframe temperatures to stabilise, called for a large fuel capacity, which resulted in a fuselage (excluding the nose probe) over 71ft long. The broad wings had a biconvex section 4 per cent thick, made in separate portions inboard and outboard of the large nacelles for the two de Havilland Gyron Junior afterburning engines, each rated at 14,000lb maximum for take-off. A particular feature was that it was stipulated that the airframe had to be made almost entirely from various steels, which greatly increased the difficulties. Eventually the first of two 188s made its first flight on 14 April 1962. I wrote a detailed description for *Flight International* on 3 May 1962, which was a little too soon to report that the 188 could not reach Mach 2, need not

The Bristol 188 looked striking, but never came near to soaking at the planned 300°C for long periods. This was the first take off in April 1962. Note the container for the anti-spin parachute at the tail. *PRM Aviation*

have been made of steel, and was going to have a very short and unproductive career. Typical, you say?

At this point it is appropriate briefly to look at two further US research aircraft which were designed to push up the Mach scale. One, the Douglas X-3, was an exact US equivalent of the Bristol 188, though ten years earlier. Intended to soak at Mach 2 for 30 minutes, the X-3 was a product of the main Douglas plant at Santa Monica. It looked fantastic, with a needle-nosed 67ft fuselage riding on a tiny wing of 22ft 8in span. I believe this wing to have been the only one with a supersonic aerofoil profile ever to have flown a manned aircraft. It was a parallel

double wedge, or flat hexagon, with a sharp leading edge and t/c of 4.5 per cent. So that the wing might lift at low speeds the entire leading edge was hinged to form a droop flap, while the trailing edge had split flaps inboard of the ailerons. Fairings on the underside covered the flap and aileron power units, which could not fit inside the wing. The engines were afterburning Westinghouse J34s, each rated for take-off at 4,850lb, fed by plain lateral inlets. Titanium was used, for the first time in primary structure, around the engines and for much of the tail, including the slab tailplanes. The pilot got aboard by being winched up on the downward-ejecting seat.

Douglas pilot Bill Bridgeman began the flight test programme on 15 October 1952. He found a reluctance to get airborne, finally leaving the ground at 260mph and flying for just a mile. Subsequently the only X-3 to fly did make 51 flights, and on one of them it actually made a contribution to knowledge. On 27 October 1954

Another disappointment was the startling Douglas X-3, understandably dubbed 'The Stiletto'. It had nothing like enough engine thrust. *PRM Aviation*

NACA pilot Joe Walker experienced complete loss of control and wild gyrations. Eventually recovering, he tried a second time to roll the aircraft by the ailerons, this time encountering even more violent pitch/yaw gyrations which helped significantly in the understanding of inertia coupling. Basically, however, the overwhelming lack of engine power made the X-3 belie its appearance. Capable of a mere 0.9 on the level, it was once dived to 1.21, but probably its greatest value lay in forcing tyre manufacturers to develop tyres that could withstand unprecedented rotational speeds.

In contrast, the Bell X-2 was quite a different animal. At first glance it looked very like an X-1 with a swept wing mounted in the low position, but it was actually a completely fresh design. Remarkably, the contract for two XS-2 (later called X-2) aircraft was signed on 3 July 1947, before the first XS-1 (X-1) had exceeded Mach 1. The objective was to use X-1 technology in a new airframe with a very much more powerful rocket engine, in order to reach new realms of speed and altitude. So fast was the X-2 expected to be that the fuselage was made largely of K-Monel, a corrosion-resistant nickel/copper alloy, while all wing and tail skins were stainless steel. This alone presented challenges, but these were compounded by the t/c ratio of 5 per cent, and the need to fit powered outboard ailerons, and inboard flaps interconnected with leading-edge droop flaps. The horizontal tailplanes were powered slabs.

The large fuselage housed more than double the load of lox and water/alcohol carried by the X-1s. As a

result the loaded weight was also doubled, to 24,910lb compared with 12,250lb for the first X-1. The engine was the Curtiss-Wright XLR25, with a lower chamber controllable up to 10,000lb sea-level thrust and an upper chamber controllable up to 5,000. Thus the pilot could control thrust smoothly from 2,500 to 15,000lb, rising at high altitude to about 21,000lb. This was a bit different from the rather primitive 6,000lb of the XLR11, especially as later in the programme the XLR25 nozzles were fitted with big extensions to expand the jets further at high altitude to reach jet velocities close to 5,000mph. Other features of the X-2 were an ejectable pilot capsule in the nose, manually retracted twin nosewheels but a main landing skid, and miniaturised instruments because there was no room for full-size ones.

Studies began in 1946. Like everyone else, Bell was apprehensive about how swept wings would behave at low speeds and high AOA. The Navy and Douglas were among the other concerned people, and to get answers the Navy turned two P-63A Kingcobras (oddly, they called them L-39s, whereas the Navy P-39 Airacobra was the F2L) over to their maker to have swept wings fitted. Bell fitted 35° swept wings, which test pilot A. M. 'Tex' Johnston (later of Boeing and then Guppy fame) found no problem. Then the wings were made more like those of the X-2 by gluing on broad sharp-edged leading edges; but they overlooked the centre of pressure shift, which posed a severe tail-heaviness problem for pilot Jack Woolams.

While the first X-2, USAF 46-674, received its awesome engine, the No. 2 aircraft, 6675, began gliding flights on 27 June 1952. The only real problem was that, like the X-1s, the X-2 slammed on to its nose gear on landing, breaking it. Number 6675 was later fitted with its engine, and on 12 May 1953 it was taken aloft under its EB-50A carrier aircraft. It exploded, and was jettisoned to destruction, the cause being later traced as yet another ulmer leather disaster'. So it was left to 6674 to do the powered flying, which did not begin

Bell X-2 46-674 pushed up the Mach scale to 3.196. In the background are the B-50 launch aircraft and support personnel. Sadly, both the Bell X-2 and Douglas X-3 suffered fatal accidents, though from totally different causes. *Philip Jarrett*

until 18 November 1955. On 23 July 1956 Pete Everest accelerated to a new record at 1,900.34mph (Mach 2.8706), the instrumentation apparently justifying such incredible accuracy. On 27 September 1956 Capt Milburn G. Apt made his first flight in the X-2, but it was no mere familiarisation sortie. He was told merely to keep within certain fairly tight bounds of trajectory, and despite his inexperience he did this precisely. The engine burned several seconds longer than expected, and after reaching 72,000ft Apt put the nose down and at burnout hit Mach 3.196, or 2,094mph. This was the first time in history that Mach 3 had been reached. Sadly, the X-2 then ran into the old problem of inertia coupling and, though Apt managed eventually to recover control, he decided to eject the nose capsule. He was still in it when it hit.

The next-generation research aircraft was the X-15. This and the stillborn D-558-III are described in the final chapter.

Having referred to the A.4 rocket in Chapter 4, it is worth briefly mentioning here that by 1957 both American and Soviet designers had created rocket-powered ballistic (wingless) vehicles that pushed up the Mach scale in a way that made winged devices look sluggish. While the A.4 reached about 3,355mph, the intermediate-range ballistic missiles (IRBMs) reached about 10,000mph, which enabled them to throw their warheads up to 1,725 miles. The intercontinental IRBMs reached about 15,500mph, for a range of up to 6,325 miles. Satellite launchers needed at least 17,600mph, and the fastest spacecraft so far, the German Helios I, launched in 1976, was accelerated by a Titan-Centaur booster to 149,129mph. But all these high speeds are attained beyond the atmosphere. Nevertheless, design of these vehicles did require a lot of new technology, including hypersonic aerodynamics and materials able to withstand severe kinetic heating in order to stand up to the conditions during the climb up through the atmosphere.

In 1955 the USAF began in earnest to fund various items intended to lead to a successor to the Lockheed U-2. This notorious aircraft had used extreme altitude to avoid interception, but for the late 1950s it was considered that high speed would also be needed. Attention focused on the Lockheed CL-400. This had a dart-like fuselage over 160ft long, with a stubby wing amidships. On each wingtip was to be a Pratt & Whitney 304-2 turbojet burning liquid hydrogen. The 304-2 is described in another of my books for PSL, *World*

Encyclopedia of Aero Engines. The problem was the low density of liquid hydrogen. Even with the CL-400's huge fuselage tanks the radius of action was only 1,250 miles, cruising at Mach 2.5 at 99,500ft. So in 1958 designer Kelly Johnson turned back to hydrocarbon fuel. Out of his studies came the fastest aircraft ever put into regular operation anywhere.

On 29 August 1959 Lockheed's A-12 was selected from five proposals, and a year later the CIA ordered 12, the contract and the funding being hidden in other appropriations. The first A-12, powered by two interim J75 engines, made a frightening 'hop' on 24 April 1962, and then a proper flight, with the stability-augmentation system operative, two days later. From the A-12 were derived the YF-12A missile-armed interceptor and the bigger and heavier SR-71 reconnaissance aircraft.

These aircraft, the so-called Blackbird family, are famous around the world. Everything about them was a challenge. The development of the airframe, made mostly of B-120 titanium alloy, the systems, designed to soak at up to 570°C, and the Pratt & Whitney JT11D-20 'leaky turbojet', all thrust forward into totally unknown

The Lockheed SR-71A Blackbird, 64-17961, seen from a JP-7 tanker. *PRM Aviation*

regions and needed an unprecedented engineering effort. One of my favourites among dozens of anecdotes is the supplier who came up with the perfect hydraulic fluid for use at Blackbird hot-soak temperatures. There was just one minor problem: at normal room temperature it was a dirty white powder! The huge fuel system, which had to remain leakproof under pressure while all the parts from which it was made expanded and contracted differently at contrasting temperatures, was eventually filled with a fuel called JP-7, carried by the specially prepared KC-135Q air-refuelling tankers. TEB (tetraethyl borane) was used to ignite both the main engine (which has the military designation of J58) and its huge afterburner. As explained in the next chapter, at over Mach 3 virtually all the propulsive thrust comes from the mighty variable inlet and nozzle, the engine itself just getting in the way.

No other aircraft have ever flown so many hours at Mach 3 or above. SR-71As still hold three absolute world records: height in sustained horizontal flight, 85,069ft; speed in a straight line, 2,193.17mph; and speed in a closed circuit, 2,092.294mph. These records were set on 27 and 28 July 1976. The speeds correspond to Mach 3.17-3.32. Aircraft enthusiasts everywhere were sad when the SR-71A operational funding was cut off in November 1989, leading to withdrawal from service by February 1990.

Although the Blackbirds were so well known, it may be worth briefly noting a few design features. Somehow I never talked to Johnson or Ben Rich about the basic configuration, but alternatives, such as engines in the fuselage or on the wingtips, were examined in some detail. The wing is almost a delta, with about 53° leading-edge taper and pronounced conical camber outboard of the engines. The prominent leading-edge strakes along the outer sides of the nacelles generate vortices which become very strong at high AOA (see Chapter 9). The t/c is basically 2.5 per cent, an amazing achievement. At the root the wing is blended into the fuselage, and in the SR-71 the leading edge is carried forward around the nose in a prominent chine. Besides probably having a stealth function, this chine exerted extremely powerful beneficial effects, at both the highest and lowest speeds. At Mach 3.2 it greatly increased lift along the forebody, helping to cancel out the rearward migration of the wing lift. At all speeds the chine improved directional (yaw)

stability, to the extent that cutting the chines back on the YF-12A interceptor to permit installation of the radar and IR sensors required the addition of ventral fins under the nacelles and a large folding fin under the fuselage. The twin vertical tails were powered rudders mounted on very short fixed stubs. In slow single-engined flight the rudder angles were startling. These surfaces were canted inwards 15° to reduce the rolling moment due to rudder deflection, and to keep them inboard of the vortices shed by the nacelle strakes.

The sheer performance of the SR-71 never ceased to amaze people (me, at any rate) right up to their final withdrawal. When they came on the scene they did not just knock a bit off the previous record; they moved the record figure into a totally different category. For example, in 1983 a Learjet 55LR set a new record for the 2,300-mile flight from Los Angeles to Washington DC. The time of 4hr 12min was considered remarkable. But on 6 March 1990 an SR-71 – 64-17972 en route to the Smithsonian National Air and Space Museum – flew between the same two cities in 64min 5sec. The mind simply boggles. It knocked off a few other interesting times along the way, such as St Louis to Cincinnati, 311 miles, in 8min 20sec (2,242mph). Once you get those shockwaves really leaning back, huge distances cease to mean very much.

Lockheed also designed a Mach 3.2 reconnaissance drone to be launched by the SR-71 A. Designated D-21A, this even more stealthy vehicle was built around another version of the Marquardt RJ43 ramjet, as used to propel the Bomarc. These tailless delta vehicles had fixed nose inlets matched to cruise speed. They were expendable, and designing for Mach 3 only is almost simple.

Only one other fighter need be mentioned here: the MiG-25. Since the mid-1950s fighters have tended to get slower rather than faster, but in 1958 the threat of the USAF B-70 resulted in the MiG bureau being ordered to design an aircraft able to shoot it down. By late 1959 the design was complete, and the first Ye-155P prototype flew in 1964. Four prototypes made a flypast at the Soviet Aviation Day show on 9 July 1967. Their appearance alone would have caused near-panic in the Pentagon, but the commentator stated that these interceptors could fly at Mach 3 and reach 30.5km, or just over 100,000ft. From then on the NATO name of 'Foxbat' caused shivers down many spines (Chapter 10).

Chapter 7

PROPULSION

To fly at supersonic speed it is necessary to use some form of jet engine. Almost all high-speed aircraft use a turbojet or turbofan, usually with an afterburner. Hypersonic aircraft have to use a rocket. In between comes the beautifully simple ramjet, but this has the severe drawback that it cannot start from rest.

Rockets are certainly the easiest installations to design or describe. They do not need air to be supplied for the combustion of their fuel, so all that is needed is to provide a thrust chamber, the propellant tankage, a turbopump or gas-pressure feed, and a control system. The chief reason they are seldom used to propel aircraft is their short endurance. Rolls-Royce has cleverly solved the problems of making a variable-cycle rocket which, during flight through the atmosphere, can burn atmospheric oxygen. This is outlined in the final chapter.

Ramjets can likewise be dismissed in a few lines. When I first met Roy Marquardt, in 1950, he looked ahead to the near future when thousands of his company's ramjets would provide propulsion for supersonic aeroplanes, as well as for missiles. It did not work out that way, though the ramjet is so obviously superior (except at take-off and landing) that many people are at last trying to produce attractive variable-cycle engines, like that designed over 50 years ago for the XF-103. These will be described later.

The turbojet is the simplest possible form of gas turbine, comprising a compressor, combustion chamber and turbine. The turbine extracts just enough energy from the flow of hot gas from the combustion chamber to drive the compressor. Downstream of the turbine, the gas is allowed to accelerate through a nozzle, convergent in a subsonic aircraft, to provide the high-speed jet that propels the aircraft. A turbofan differs in that the upstream low-pressure (LP) part of the compressor pumps more air than is needed by the rest of the engine.

This extra airflow is bypassed around the rest of the engine, straight to the jetpipe. Bypass ratio (BPR) is the ratio of air bypassed around the engine to that passing through it. For best fuel economy and low noise, the latest subsonic turbofans have BPR of 5 to 11. In the most powerful engines, this makes the LP compressor, or fan, very large, the diameter reaching 10ft or more. In contrast, supersonic aircraft need slim engines, so the BPR is unlikely to exceed 1.5, and is often only about 0.2 or 0.3, the bypass airflow being little more than that needed to keep the casing of the engine cool.

For supersonic flight, either type of engine usually needs an afterburner, or reheat. The jetpipe downstream of the turbine is enlarged and fitted with a nozzle whose profile and final area are adjustable. Downstream of the turbine are rings made of tubes which can spray fuel into the hot gas, and immediately downstream are gutter rings or other devices which locally cause turbulence for reversal of the gas flow. For supersonic flight, fuel is pumped to the spray rings, ignited, and allowed to burn in the turbulence downstream of the gutters. Without the gutters the extra fuel would be swept out of the jetpipe nozzle before much of it could burn. In the engine upstream, whether turbojet or turbofan, the highly stressed turbine blades impose a clear limit on the maximum gas temperature. In the afterburner there is no such limitation, and the gas can leave the nozzle at 1,900°C or even more. Such a flow of intensely hot gas can be accelerated to highly supersonic speed by letting it escape through a con/di nozzle. We have already seen how a divergent nozzle is needed to accelerate a supersonic flow. Downstream of the nozzle the jet can be seen to contain a series of brilliant golden 'shock diamonds', which are caused by shockwaves within the jet being repeatedly reflected from its edges. The shearing action at the boundary between the jet and the atmosphere causes intense noise. At high

SUBSONIC INSTALLATION

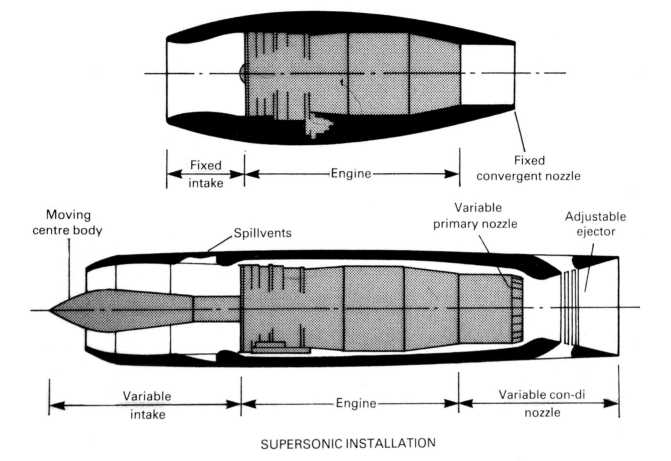

SUPERSONIC INSTALLATION

altitude the reduced atmospheric pressure makes the jet expand into a large plume, which quickly eliminates visible shock diamonds.

Early air-breathing (i.e. non-rocket) supersonic aircraft used turbojets. This type of engine was used by Concorde, and is still found on a few military aircraft. All of the latest fighters and bombers have turbofans. The main difference with these engines is that downstream of the turbine the jet contains much more uncombined oxygen than is the case with a turbojet, because all the outer part of the jet is atmospheric air that has not passed through the engine's combustion chamber. It follows that far more fuel can be burned in the afterburner. With a turbojet, cutting in the afterburner may boost thrust by about 30 per cent, whereas with a turbofan the gain may be 100 per cent. In effect, it doubles the size of the engine, and makes it much more suitable for supersonic flight by greatly increasing the jet velocity, but at the cost of extremely high fuel

These purely conceptual diagrams outline the great difference between the simple installation of a turbojet or turbofan in a subsonic aircraft and the longer and more complicated installation needed for supersonic flight. The one thing that does not increase is frontal area.

consumption. With a modern turbofan, the increase in thrust between Military (or maximum cold) thrust and full afterburner may be 100 per cent, while the increase in rate of fuel burn may be 600 per cent.

Another name for an advanced supersonic engine (turbojet or turbofan) with an afterburner is a turboramjet, sometimes written with a hyphen. We have already seen in the XF-103 how it is possible to make an even cleverer form of turboramjet in which, at supersonic speed, the

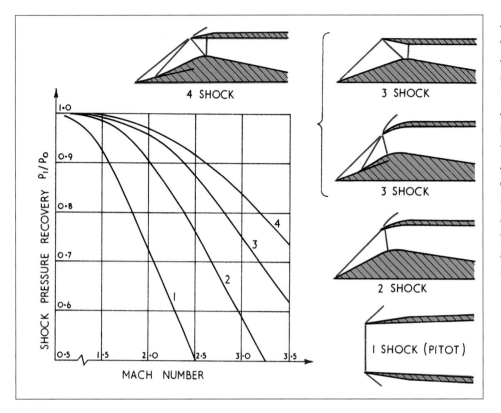

A simple (but sharp-edged) inlet is adequate for Mach numbers up to about 1.5, though many aircraft, including the F-16 and F/A-18, use them at up to M 2. At M 2 such an inlet cannot have a pressure recovery greater than about 72 per cent, as the graph shows. Fast aircraft therefore use multi-shock inlets.

basic engine is bypassed entirely, the air going direct from the aircraft inlet to the ramjet, which at take-off was the engine's afterburner. We shall return to such powerplants, which are called variable-cycle (VC) engines, later. One difficulty with ramjets and afterburners is that they burn fuel so fast. Until the 1980s it was taken for granted that supersonic flight meant 'going into 'burner', but the newest USAF fighter, the Lockheed F-22A Raptor, has been designed as a 'supercruise' aircraft. This means it is designed to cruise at supersonic speed for long periods, without using afterburner.

Before looking at today's and tomorrow's supersonic engines, it is essential to recall that, in almost every supersonic aircraft, the engine is installed between a carefully designed variable inlet and an equally carefully designed variable nozzle. The inlets and nozzles are among the most difficult parts of any supersonic aircraft to design, or to construct. Aircraft were simple when they took off at 40mph and cruised at 80mph When they take off at 180mph at sea level, and cruise at 2,000mph at over 70,000ft, the problems are harder. You really need two totally different flying machines, not only with contrasting forms of inlet, engine and nozzle, but also with different wings and flight-control systems. In the

course of each flight the variation in air density might be 20:1, the variation in ram pressure inside the inlet might be 36:1, and the difference in stagnation temperature (before the air even enters the engine) well over 500°C.

A little was said about inlets in previous chapters. Clearly, the simplest inlet is a plain hole, a so-called direct pitot (pronounced pee-toe) inlet. It can be in the nose, on the sides or underneath, and the only stipulation for supersonic flight is that its edges have to be quite sharp, whereas subsonic inlets are well rounded. When the aircraft accelerates, as the Mach number reaches 1 a normal shock forms ahead of this type of inlet, like putting cling-film over a dish. At perhaps Mach 1.4 the shock will be essentially in contact with the lip all round. As the air passes through this shock it is almost instantaneously raised to a much higher pressure, which is further increased in the expanding subsonic diffuser downstream. The loss in total pressure at the shock is insignificant at low supersonic Mach numbers. This loss is only 10 per cent ($P_1/P_0 = 0.9$) at Mach 1.6. Accordingly the F-16, F/A-18 and *Rafale* (for example) have simple fixed inlets. But if the design peak Mach number is higher than 2, a plain pitot inlet is no longer adequate. At Mach 2 the pressure loss is 28 per cent, and

Supersonic engines can suffer from severe inlet problems whenever conditions depart from point B.

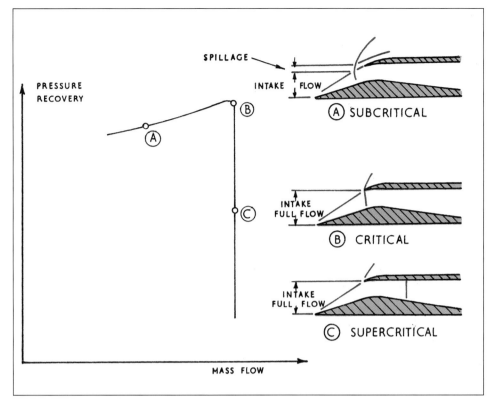

So far the reader has probably interpreted these inlets as circular, but of course they can be of different form. The F-104 and Mirage III pioneered semi-circular lateral inlets with half-cone centrebodies. The F-4 Phantom II is one of many aircraft with lateral inlets of two-dimensional form, in which the diagram of the centrebody and outer wall represents a horizontal section. The A-5 Vigilante and MiG-25 were among the first aircraft with two-dimensional inlets which in effect are rotated through 90°, so that the 'centrebody' becomes the upper wall, and the lip the lower wall. In this case the diagram shows a section through the inlet in the vertical plane. A glance at a supersonic aircraft shows which type of inlet has been used. Among those with the F-4 arrangement are the B-70, Su-15, Su-24 and Tu-160, while the A-5 configuration is seen on the F-14, F-15, MiG-29, Su-27, Concorde and the Tu-22-M3.

at 2.5 (peak level Mach for the F-15, for example) no less than 50 per cent.

Accordingly we have to insert a sloping surface ahead of the intake lip to create an oblique shock. This makes the normal shock weaker, and reduces the loss in total pressure. There are many ways in which we can arrange the geometry of the sloping surface and the lip. If the inlet is in the nose, it is logical for it to be circular, and for the sloping surface to be provided by a conical centrebody, as in the Lightning or MiG-21. For flight at higher Mach numbers, the centrebody needs to be relatively larger, and the surrounding gap for the airflow to be smaller, as it was in the SR-71. By making the central cone in sections, with progressively steeper angles, it is possible to precede the normal shock by two or more oblique shocks. This is called an Oswatitsch inlet, and a good example can be seen on the MiG Ye-152-1 (publicised as the 'Ye-166') which set speed records in 1961. Each change in centrebody angle creates another oblique shock. The diagram opposite shows the beneficial effect of a three-or four-shock inlet, though the designer has to work hard to avoid flow breakaway and turbulence where the inclined shocks strike the boundary layer over the centrebody.

The F-111 used an unusual quarter-round inlet tucked under the wing root and, partly because it was too near the front of the engine, the inlet suffered severe problems (Chapter 10). Eventually a bigger and improved inlet was made to work, but the experience drove home yet again the vital need to match the inlet to the engine. What makes this task harder is that a simple fixed-geometry

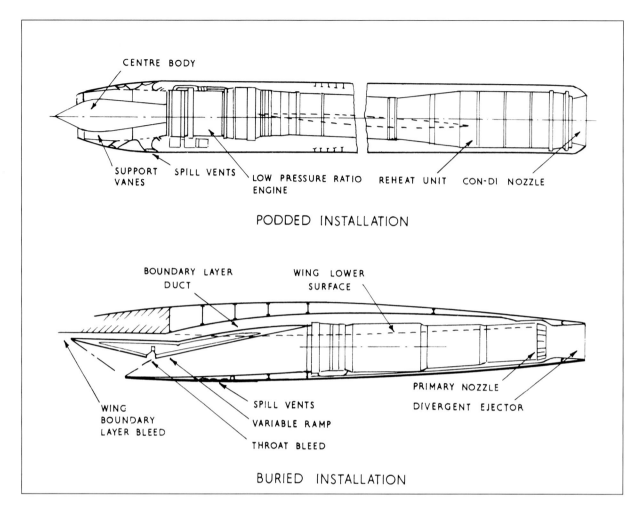

CENTRE BODY

SUPPORT VANES   SPILL VENTS   LOW PRESSURE RATIO ENGINE   REHEAT UNIT   CON-DI NOZZLE

PODDED INSTALLATION

BOUNDARY LAYER DUCT   WING LOWER SURFACE

WING BOUNDARY LAYER BLEED   SPILL VENTS   VARIABLE RAMP   THROAT BLEED   PRIMARY NOZZLE   DIVERGENT EJECTOR

BURIED INSTALLATION

Supersonic engines are usually installed either in a pod, with a circular inlet, or buried in a fuselage or underwing nacelle. The buried installation appears to be a side view, but looking down on an Su-24 engine (for example) would give a similar picture.

inlet is matched to the engine only at one particular design condition. This condition may be take-off, or Mach 1 at sea level, or Mach 2 at 35,000ft and above. Whichever of these conditions is chosen, the inlet will be inefficient – perhaps grossly so – at the other two. In supersonic aircraft the inlet is almost always designed to a particular mass flow, with the first (or only) oblique shock exactly crossing or touching the lip, and the normal shock at the diffuser throat. This is shown as point B on the accompanying figure (see page 99), which depicts a typical inlet with a straight-sided cone or wedge resulting in a two-shock system, one oblique and one normal.

Suppose the pilot of a supersonic aircraft throttles back. The mass flow will fall, and the normal shock will be expelled forward, out of the inlet, to allow the large flow of excess air to be spilt around the lip. The air entering the outer part of the inlet passes through a single shock formed by the fusion of the normal and oblique shocks, so pressure recovery falls (point A on the plot). This is called subcritical operation. Suppose instead that the pilot pushes the throttle all the way forward, to get maximum power. This increases engine speed, sucking harder at the air in the inlet duct, and the reduced pressure pulls the normal shock away from the throat and into the diffuser, gaining strength as the Mach number in front of it increases. But the air entering the inlet knows nothing of this, because the flow is supersonic, so mass flow cannot increase and, instead, pressure recovery falls dramatically (point C).

So, in practice, almost all supersonic inlets are made with variable geometry. To keep this book to an affordable size I won't go into much detail, except to comment that

Believe it or not, these sketches of the SR-71 nacelle are considerably simplified. At cruising speed, around M 3.2, the main spike shock is focussed on the lip all round.

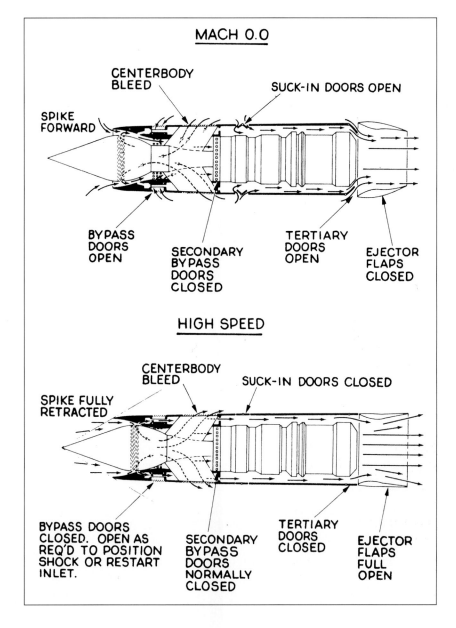

the complexity of the inlet naturally depends on the mission of the aircraft. The greater the range of Mach number, and the longer the mission radius, the more sophisticated the inlet system must be. An SST (Chapter 9) is designed to cruise at almost constant Mach, and over quite a small range of altitudes, yet it still needs a complex variable inlet and nozzle system in order to take off in a reasonable distance, and with acceptable noise, and still match the cruise regime. In each case there are bound to be either variable ramps or a translating (sliding in and out) centrebody, auxiliary inlet doors, and overboard spill vents or doors, as well as various auxiliary bleeds, especially at the throat. Except for nose or SR-71-

type podded inlets, there will also have to be some way of diverting the sluggish and turbulent boundary layer, which flows back along the skin upstream of the inlet. Sometimes this is quickly dumped overboard, while in other aircraft it is ducted back to cool the afterburner skin and nozzle.

One diagram (opposite page) shows in simplified form a two-dimensional inlet such as that used on Concorde or the Su-27. If one replaces 'wing lower surface' by 'side of fuselage', then it could apply to an Su-24 or F-4. In many such installations the lower (or outer) wall and lip are fixed; on others at least the lip is hinged, one example being the Typhoon. In every case the other side

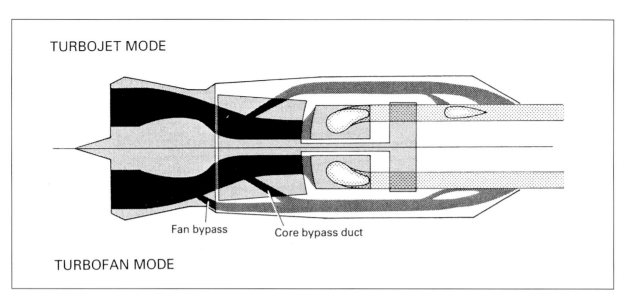

TURBOJET MODE

Fan bypass   Core bypass duct

TURBOFAN MODE

In 1990 General Electric released this highly stylised diagram showing the two operating modes of the F120 variable-cycle engine developed for advanced fighters.

of the inlet, the upper or inner wall, is pivoted so that, as Mach number increases well beyond 1, the throat can be made progressively smaller. At the same time the nozzle is altered from convergent to con/di, until at really high Mach numbers, of the order of 3, the nozzle exit is much larger in cross-section than any other part of the installation. I described and illustrated hypothetical Mach 3 installations in some detail in *Flight* for 11 May 1956, and things haven't changed much since. But not many people have actually flown such installations, and I am grateful to Kelly Johnson and Lockheed for two simplified diagrams of a typical 'Blackbird' installation (see page 101).

These diagrams ignore the fact that the J58 was a pioneer variable-cycle (VC) engine. At take-off it was a single-spool turbojet, but near Mach 2 valves opened to bleed an enormous airflow from the fourth stage of the compressor, piping this back to re-enter the engine upstream of the afterburner. This increased compressor stall tolerance, and provided air to cool the afterburner and burn more reheat fuel. When an SR-71 took off, the thrust came from the engine. Some parts pulled forward while others pushed back, and the resultant thrust, the difference between two much greater forces, was

imparted through the engine pick-up points. At Mach 3 the situation was totally different. Starting from the front, the huge spike caused drag, a force to the rear, of 14 per cent of the total. Next came the inlet diffuser, from the throat to the face of the compressor, and this generated 70 per cent of the thrust. The engine generated a mere 17 per cent, and the ejector nozzle 27 per cent. Adding 17 + 70 + 27 we get 114 per cent, and taking off the spike drag we are left with 100 per cent. Designer Johnson used to say, 'Pratt's J58 is only a flow inducer; it is the nacelle that pushes the airplane!'

We have already encountered a turboramjet in the XF-103, and have seen that GE's first supersonic turbofan fighter engine, the YF120, was of the variable-cycle type. I cannot close this chapter without a brief further discussion of VC engines, because the F120 is likely to set the pattern for future fighter engines for as far ahead as can be seen. In 1990 I had a long talk with Brian Brimelow, manager of the programme. His YF120 was one of the candidate engines for the ATF (Advanced Tactical Fighter, Chapter 11). He was very convincing in stressing that such aircraft need a VC engine. For take-off, climb and subsonic loiter you need a turbofan. Brimelow explained: 'The fan blades are huge. The YF120 is a very high-pressure engine, and this, together with the optimised bypass ratio, gives the lowest possible fuel burn in subsonic flight. For dry supercruise we close the fan bypass doors, virtually turning the engine into a turbojet. We leave just a small flow through the core bypass duct for back-end cooling, and to provide extra air should we wish to afterburn. Of course, we have to be very clever to preserve stealth features at the inlet and nozzle.'

Rivals would obviously suggest that a VC engine must be complicated, but Brimelow claims, 'In fact, the reverse is true. Thanks to the VC design the YF120 is mechanically simple. For example, we have only half as many stages of turbine blading as in the Fl10. There has to be a ring of nozzle guide vanes controlling the flow of hot gas from the combustor. Then we have just a single-stage HP turbine rotor and a contra-rotating single-stage LP rotor. That's it, nothing else.'

On the same occasion, Pratt & Whitney's William C. Missimer, President of Government Engine Business, told me, 'I wish you could talk to our Brimelow, but Walt Bylciw [pronounced Bilsher] is out in the desert looking after the YF119 engines in the first YF-23 Advanced Tactical Fighter (ATF). You will appreciate we pioneered the VC engine in the J58 of the Blackbirds. We understand the propulsion requirements of the ATF very well, and the USAF gave us a free hand to design the Fl19 any way we wished. In our opinion we have devised a simpler, lighter solution with a fixed-geometry turbofan. Of course, it has a clever two-dimensional nozzle, with variable profiles and inflight vectoring.' Both the rival ATF engines established a level of thrust/weight ratio previously achieved only in much simpler VTOL lift engines. Who won? See the end of Chapter 11.

One of the VC species, much bigger than the ATF engines, will eventually be picked for the second-generation SST. Of course, the Olympus 593 of Concorde had a fully variable inlet and nozzle, all tied together by the first electronic control system ever to go into routine operation. This engine, the roots of which go back to 1946, logged over half a million hours at Mach 2 or above, which is many times more than all the military engines of the Western world combined. Concorde flight time was 243,845 hours.

Despite the shrill objectors to Concorde, it seems unthinkable that this pioneer journey-shrinker should have no successor. Moreover, as explained in Chapter 9, today we can do very much better. Virtually all of the big engine design teams, in Britain, the USA, Japan, France and the former Soviet states, have been quietly studying how best to propel 'SST2'. There is much unanimity, and it would be logical for an international programme to be assembled in due course. Indeed, in the previous century SST2 often seemed quite close, whereas by 2008 it had receded.

A future SST propulsion system must meet the noise legislation of ICAO Annex 16 Chapter 3, which among other things demands take-off noises 15 to 20 Equivalent Perceived Noise Decibels (EPNdB) lower than Concorde. To meet the sideline-noise requirement, the velocity of the mixed jet issuing from the nozzle cannot exceed about 1,500ft/sec, assuming a 680,000lb aircraft with four 60,000lb-thrust engines. This is very much less than the

Rolls-Royce's suggestion for a tandem fan VC engine for future SSTs. Rather similar technology is likely to be needed for future V/STOL engines.

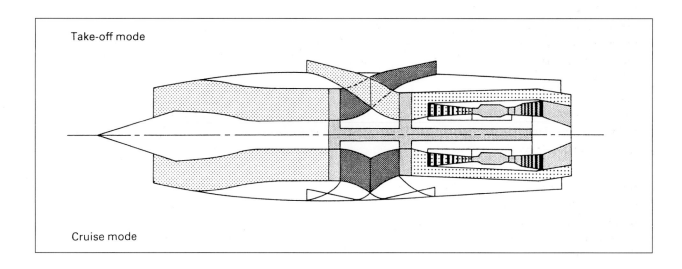

Take-off mode

Cruise mode

jet velocity of today's afterburning engines. A modern augmented turbofan designed for optimum efficiency, cruising at high altitude at Mach 2, would have a take-off jet velocity of about 3,000/ft sec, while the engine for a Mach 3 aircraft would require 3,800ft/sec. Either would be deafening. The only answer is a VC engine.

To make a VC engine, space must be provided for the large and strong movable features, and their drive systems. The photograph of the Soviet SST engine shows that, like the Olympus 593, there is some room around its wasp-waisted middle, but not enough. At Mach 2 the area of the intake stream-tube (the 'pipe' of air flowing into the inlet) is almost exactly the same as the face of the compressor, so a Mach 2 VC engine would have to be bulged amidships, increasing the overall frontal area, which for a supersonic aircraft is undesirable. At Mach 3, however, the intake must be about 1.6 times the area of the entry to the engine, and this does provide room for VC features to be hidden behind the front of the engine.

VC engines have many features in common with those for powered-lift, short take-off and vertical landing (STOVL), such as the tandem-fan engine. Indeed, tandem-fan engines are seen by Rolls-Royce as among the best for both applications, at least for Mach 2. As a simple drawing shows, at take-off all the air entering the main inlet is accelerated by the front LP fan, and then discharged at once through a ring of nozzles. Between these nozzles are auxiliary inlets which admit air to the rear LP fan, which feeds the engine core. Typical data for a Mach 2 SST engine of this type would include a mass flow, at Mach 0.3 at sea level, of 942.7lb/sec, and a thrust of 51,500lb. This airflow is 1.9 times that of an equivalent conventional engine, which is very close to the value needed to meet noise legislation. The lower half of the drawing shows that, at Mach 2 cruise, the auxiliary inlets and nozzles are retracted, and doors are opened so that all the air enters at the main inlet and passes through both fans, and then through or around the core. The variable nozzle is not shown. Subsonic cruise data for Mach 0.93 at 36,090ft would include a mass flow of 334.3lb/sec, thrust of 10,000lb and s.f.c. (specific fuel consumption) of 0.83. Figures for Mach 2 at 53,000ft would include a mass flow of 260.6lb/sec, thrust of 14,520lb and sfc of 1.181. These figures are equal, or even superior, to those of any fixed-geometry engine. As with Concorde, the subsonic cruise sfc would apply when flying over land. Powerplant weight should be similar to

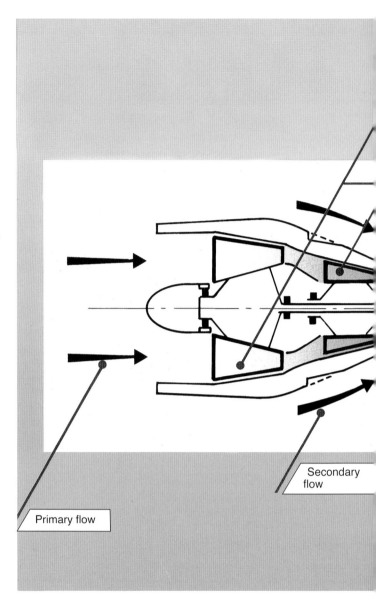

Primary flow

Secondary flow

that of a conventional engine, and the variable features could be stowed within the area of the inlet.

In 1990 Rolls-Royce published a drawing, not suitable for reproduction, contrasting the materials used in Concorde's Olympus 593 turbojet with those selected for a second-generation SST engine. Basically, the Olympus engine used only two materials: the LP compressor and the first three high-pressure (HP) stages were of titanium alloys (450°-650°C), while everything downstream was of nickel-based alloys (discs 550°-725°C, blades 950°-1,150°C). Tomorrow's engine would use the Ti alloys only for the first two LP stages. Next would come a section of structural composites, also used for the entire

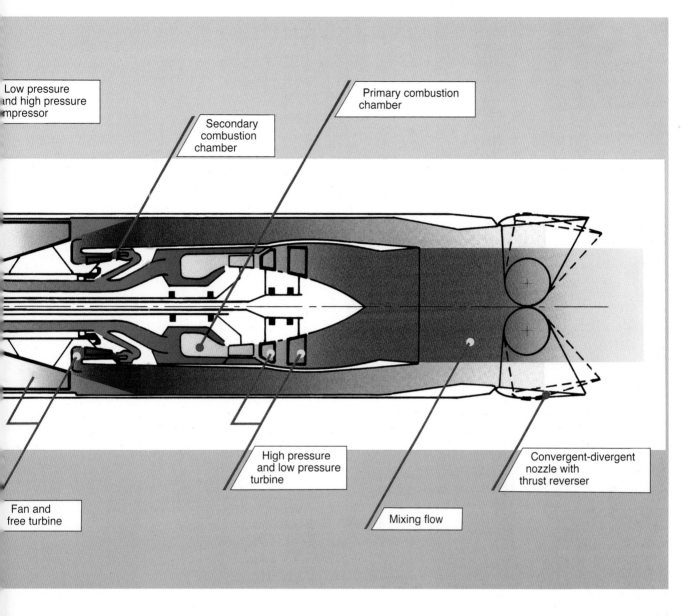

Low pressure and high pressure compressor

Secondary combustion chamber

Primary combustion chamber

High pressure and low pressure turbine

Convergent-divergent nozzle with thrust reverser

Fan and free turbine

Mixing flow

bypass duct (300°C). All the rest of the LP and HP spools would be of reinforced titanium aluminides (800°C). The turbines and shafts would be of Ni-based alloys, and the combustion chamber, turbine stators and core jetpipe would all be of composite ceramics (1,400°C).

In late 1989 Rolls-Royce signed an agreement with Snecma, their French partner on the Concorde engine, to carry out a joint study programme on new SST engines. At the same time Snecma published the configuration adopted in its own MCV 99 (*moteur cycle variable*), which was a variation on the tandem-fan formula. For take-off and subsonic operation the MCV 99 would be a turbojet, with an additional bypass flow induced

With true Gallic boldness, SNECMA propose quite a complex SST engine, designated MCV99. It is shown here in the take-off mode, with secondary flow through a row of side inlets, pumped by a fan driven by what amounts to a second engine, which is shut off in cruising flight.

through lateral or ventral nacelle inlets by a free-running fan driven by an adjacent turbine taking gas from an auxiliary annular reverse-flow combustion chamber. This arrangement would give low noise and minimum fuel burn. In cruising flight the fan inlets would be closed, the secondary combustor fuel stopped and the fan brought to rest. All flow would then go through the core, matching the engine to Mach 2-2.5, without afterburning. MCV 99 has been studied with various numbers of turbine, compressor and fan stages, some having two-spool compressors. At the 1990 Farnborough show a brochure showed a two-spool engine with a two-stage fan, while large artwork showed a single-spool 10-stage compressor and a large single-stage fan. Much has been done since to refine the concept.

Both Rolls-Royce and Snecma have expressed interest in participating in a Japanese programme funded by the Ministry of International Trade and Industry. As noted in Chapter 9, Rolls-Royce has collaborated with Lyul'ka, now part of Lyul'ka-Saturn, on a supersonic bizjet engine, while Snecma has held discussions with the Tupolev bureau regarding propulsion of a large 250/300 seat SST to succeed the Concorde-rival Tu-144.

For cruising at Mach 3 Rolls-Royce would propose an aft-fan design, and in some ways this is simpler. At take-off the inlet, core nozzle and final nozzle are all wide open. A free turbine added behind the main engine thus extracts high power to drive the big aft fan, to pump almost double the core airflow. Data for Mach 0.3 (take-off and initial climb) at sea level: mass flow 1,192lb/sec; thrust 62,716lb. As the aircraft climbs and accelerates, a ring of front doors progressively close to reduce the fan airflow, while the core nozzle slowly closes to reduce the power extracted by the free fan turbine. Data for subsonic cruise at Mach 0.93 at 36,090ft: mass flow 447lb/sec; thrust 12,000lb; s.f.c. 0.838. At the same height, figures for transonic acceleration at Mach 1.2 would be, mass flow 664lb/sec; thrust 32,353lb. At Mach 3 cruise at 70,000ft only 10 per cent of the air can pass through the fan, and the power taken out by the turbine is correspondingly reduced by the small core nozzle, but the final nozzle is changed to the required con/di form, to give: mass flow (total) 346.3lb/sec; thrust 16,250lb; s.f.c. 1.454. On paper this engine looks impressive, but more work is needed to increase take-off airflow to the ×2.5 target for low noise. Alternatively, the noise limit could be met with a small cruise penalty. But, perhaps astonishingly, we don't appear likely to need a Mach 3 airline engine for many years.

Projected supersonic engines typically assume a compressor delivery temperature of about 652°C and an HP turbine entry temperature of about 1,377°C. These are values already exceeded by some of today's engines, both military and civil. Of course, pollution of the stratosphere is highly undesirable. Leaving a vapour trail of pure ice is no problem, but NO_x (the generalised term for all the various oxides of nitrogen produced by burning hydrocarbon fuels) emissions have to be reduced to an absolute minimum. Extremely good reductions are being achieved with subsonic engines, and the research carries across directly to supersonic aircraft.

Chapter 8

SUPERSONIC BOMBERS

Visiting Convair at San Diego in 1955, I was told by Dick Sebold, VP Engineering, 'There's only one thing more difficult than a supersonic fighter, and that's a supersonic bomber (SSB). For a long time it seemed to be impossible.' Then, off the record, he showed me pictures of the mock-up of the B-58, which was being created at the Fort Worth plant. This pioneer aircraft first flew on 11 November 1956. In many ways it was the most advanced aircraft then flying, but in fact it had been preceded by two rather simpler aircraft in the USSR which were the first SSBs to fly.

Both were made possible by the Lyul'ka AL-7 turbojet, which had a supersonic first stage and a dry rating of 14,330lb thrust. The Ilyushin Il-54 had some points of similarity to the B-47, though it was much smaller. The bicycle landing gears had an amazingly long wheelbase, so thick doubler plates were riveted outside the fuselage to stop it from sagging. The two engine pods were slung close under the high wing, whose sweep was no less than 55°. First flown by the famed Vladimir Kokkinaki on 3 April 1955, the first Il-54 was later fitted with the AL-7F, rated at 19,840lb in full afterburner, giving marginally over Mach 1 over a wide range of altitudes. Tupolev's rival, the Tu-98, was later in timing, making its first flight in June 1956. This had the engines in the rear fuselage, fed from plain inlets aft of the cockpit via large ducts passing over the thin mid-mounted swept wing. Bogie main gears of rather narrow track retracted into the fuselage. The jetpipes precluded installation of a tail turret, as in the Il-54, but twin 23mm cannon were installed in a remotely controlled barbette higher up at the base of the rudder. The published maximum speed was 848mph, Mach 1.29. Wisely, the VVS (air force) waited for the Tu-105, described later.

In contrast, the Convair B-58A Hustler went into production and for a short period equipped two wings of USAF Strategic Air Command. There was very little wrong with this impressive aircraft, which did all it was supposed to do, but it was considered inordinately costly to operate, and suffered from limited mission radius. But there are two ways of looking at most things, and in fact Convair achieved a combination of range and sustained speed never previously even approached.

It was in March 1949 that the USAF held a competition to establish the feasibility of a manned supersonic attack system. By this time there was no doubt that it would be possible to build an SSB that could carry heavy loads and operate from normal airfields, provided that it did not have to go anywhere. What made it very difficult, or impossible, was that the targets were all thousands of miles away. As explained in the next chapter, the L/D (lift divided by drag) ratio falls dramatically during transonic acceleration, so that in round figures the mission radius of an SSB is less than half that of an equivalent subsonic aircraft. This is to some degree carved in stone, like the laws of Newton, and not very much can be done about it.

It follows that the early SSB designers had to be particularly clever, and if possible 'cheat' in some way to defeat the impossible figures. In the 1950s US designers planned two SSBs with range long enough to be of interest to Strategic Air Command (SAC), and both design teams started out by cheating. One team created the XB-70 (see later). In Convair's case they put the bombload and much of the fuel *outside the aircraft*, so that on the journey home the B-58 had become much smaller. Apart from this, it was mainly a matter of using the same Lippisch-inspired tailless delta layout as for the F-102, but with the challenging t/c ratios of 4.08 at the pointed tip and 3.46 at the broad root, and hanging on

First flown in 1956, the Convair B-58 Hustler was the
pioneer supersonic bomber. It carried its load in a giant
pod.

the wing four engines of a completely new type housed
in the world's first truly supersonic nacelles. Having just
gone through the mill developing the giant B-36, the
Fort Worth team were dismayed to find they needed more
tunnel hours than the entire B-36 programme just to find
out where to put the B-58's engines. Over two million
man-hours were devoted to deciding what materials to
use in the airframe and how to fashion them. Most of the
exterior ended up as honeycomb sandwich, with cores
of glass-fibre or metal. Areas subject to severe heating
were skinned in stainless-steel honeycomb sandwich,
assembled by brazing. An important example was the

huge trailing-edge elevons, which had a chord of 7ft and
were driven by power units with the staggering hinge
moment of 120,000lb/ft (I believe the previous record,
on the B-36 rudder, was 31,000lb/ft).

One could go on and on about the challenges and
breakthroughs of the B-58. General Electric played a
pivotal role with their J79 engine. Originally Convair
had planned to use four J57s, but the J79 was not only the
world's first turbojet designed for Mach 2, but it was also
2,000lb (40 per cent) lighter. Its chief innovation was to
use a single axial compressor with 17 stages in order to
achieve the required high pressure ratios, and then make
it work by using variable-incidence stator blades ahead
of the first seven stages of rotor blading. At the back was
an advanced afterburner. For the B-58, this radical new
engine was installed inside a beautiful nacelle with a
sharp-edged inlet with a movable centrebody leading to
an unprecedented arrangement of dump valves and bleed
doors. Convair, like GE, were hoping to reach Mach 2,

but in 1953 nobody had ever got anywhere near this, except by using a rocket. The B-58 propulsion system beat Mach 2, and was pioneering all the way.

As for the rest of the aircraft, this began with a fuselage that looked unnaturally thin, because of the use of the gigantic external pod to carry most of the payload. Then, when the area rule was discovered, the body got even thinner. There were none of the familiar gun turrets, though a 20mm Vulcan cannon was put like a wasp sting in the tail, with a limited remotely controlled arc of fire. The crew of three sat in tandem in unique cockpits which, in emergency, could each be sealed inside an ejectable capsule which could subsequently serve as a shelter on land or water. The problems of making such a system work, despite the mass of flight-control rods, cable looms and other services passing through it, may in part be imagined. The job was done by Stanley Aviation, who were also making the thin wings of the X-1E. In theory, a B-58 crew would have been justified in flying in shirtsleeves, with no partial-pressure suit, Mae West, parachute or exposure suit. The navigation and bombing avionics comprised a vast system incorporating radar, doppler, star tracker, inertial platform, two radar altimeters and a complex analogue computer to tie the parts together. The landing gears had to be exceptionally tall, each comprising a double gatefold. The main gears folded forward into boxes which projected above and below the very thin wing. Despite this the available depth was only 22in, and so eight tyres were needed on each leg. The tyre inflation pressure and rotational speed exceeded anything previously attempted, the limiting speed at touchdown being 306mph. In a normal take-off the B-58 became airborne at about 255mph.

Altogether 116 B-58As and TB-58 trainers were built, and in service with the 43rd and 305th Bomb Wings they captured a shoal of world records. Most were intercontinental flights with air refuelling, while one (flown by Convair's Mr Beryl Erickson) went from Fort Worth, Texas, to Edwards Air Force Base, California, 'faster than a pistol bullet at 500ft or below'. One of the more unusual records gained a trophy donated to the *Aéro Club de France* by Louis Blériot in 1930 for the first pilot to exceed 2,000km/hr for 30 minutes. In 1930 this was pure science fiction, but on 10 May 1961 Maj Elmer E. Murphy and his crew merely set up the avionics to describe a perfect circle of the required size and then went into full 'burner. After the exact half-hour the circle was

complete, at 1,302mph (2,095km/h). Flying the B-58, if all went well, was a dream. The landing approach seemed to be on rails, yet in fact it was a demanding aircraft. Another Fort Worth pilot, Dick Johnson, was asked: 'Can you land the F-111 with the wings jammed at maximum sweep?' He replied 'Just about; it would be like landing a B-58.'

A little later twin J79 engines were picked by the Navy for the North American A3J-1 Vigilante. This carrier-based bomber, first flown on 31 August 1958, never captured the attention of the media as did other J79-powered aircraft, such as the B-58, F-104 and F-4, but I believe it incorporated more radical new design features than any other aircraft since the Wrights. Some, such as the complex lateral-control system involving giant ducts being created through the wings, appeared questionable. But the overall fuselage shape, with a long slender forward section disappearing between two giant engine ducts, is seen today in the F-14, F-15, MiG-29, Su-27 and many other aircraft. Other features included blown flaps, a 3.5 per cent wing, slab tailplanes and even a slab vertical tail with no fin (the original design had twin verticals). Between the engines was a giant tunnel from which, on an operational mission, would be shot a unique 'train' over 30ft long comprising a nuclear weapon stabilised by two empty fuel tanks attached in tandem behind it. Most Vigilantes were RA-5C reconnaissance aircraft, the first to go into service with an SLAR (sideways-looking airborne radar). During early tests, imagery from an RA-5C showed an American football field from a range of 60 miles, taken at 40,000ft at almost Mach 2. The image showed the field to be 100.1 yards in length. The field itself was then measured, and found to be 0.1 yard (3.6in) too long.

Back in February 1954 the British Ministry of Supply issued a farsighted requirement, R.156T, for a supersonic reconnaissance aircraft. At this time not only was no British supersonic military aircraft on order, but not even a supersonic research aircraft had flown. Accordingly the design teams that sought to win what looked like a potentially very important contract had not much to go on except common sense and what they could learn from the Germans and Americans. In the end three companies made bids: all three looked plausible and indeed impressive. Notable features were multiple afterburning turbojets, large fuselages and canards – at that time absolutely novel. All made provision for an internal bombload, and none adopted any radical

The NAGPAW (North American General Purpose Attack Weapon) became the A3J and later the A-5 Vigilante. When the mock-up was built in 1957 it had twin powered rudders, a feature now in fashion.

cheating methods such as jettisoning large parts of the aircraft during the mission. The Vickers 725 was the biggest, and one of its many notable features was that the titanium flaps were to be lowered into the jets from 12 RB.121 engines (operating in cold thrust). Handley Page picked the same engine for the HP100, but used only six, four Concorde-style under the wing and two

under the forward fuselage. But the preferred choice was the Avro 730.

Avro submitted their proposal in May 1955 and received a contract in September for ten development aircraft, preceded by two small-scale Avro 731 research aircraft. Studying this programme from over 50 years later, I am struck by the enormous amount of thoughtful detail engineering that went into it, in many cases providing answers to problems that the British industry has never addressed since. For example, nowhere in Britain in the past 53 years has anyone been in detail design on an aircraft intended to cruise at Mach 2.6 for 5,450 miles. These figures applied to the original aircraft with two superimposed Armstrong Siddeley P.159 engines on each wingtip. In 1956 the design was altered to meet the revised RB.156D specification, for a reconnaissance bomber. Apart from confirming the internal weapon bay, or alternatively a 50ft 18,000lb thermonuclear bomb recessed under the fuselage, the final aircraft had modified wings exactly similar in profile and plan to those of the later Blackbird series, with the nacelles each housing four P. 176 engines. Armstrong Siddeley's H. S. Rainbow said this was 'by far the best engine we never tested'. Span, originally 59ft 9in, became 65ft 7in, and length was reduced from 163ft 6in to 159ft.

Originally the wing had been tapered equally on both the leading and trailing edges, but in the revised form the taper was nearly all on the leading edge, the angle outboard of the nacelles being 64°. The t/c ratio remained 3 per cent, and the rectangular canard, with powered elevators, also remained unchanged. Avro demonstrated that the flapped canard was the optimum way of trimming out the aft movement of CP during transonic acceleration. The all-moving foreplane also greatly increased lift at take-off and landing, and avoided the potentially severe problems of a horizontal tail, which would in any case have pushed downwards instead of upwards on take-off and landing. Yet a further advantage was that it enabled the giant Red Drover SLAR to be installed between the foreplane and wing without electronic interference. The landing gear comprised a main unit close behind the CG, a steerable nose unit based on that of the Vulcan, and stabilizing outriggers under each nacelle. The main unit took almost all the weight, and comprised a four-wheel bogie which sufficed for take-offs with up to 60 per cent fuel; above this weight four extra wheels were added, to be jettisoned after take-off. Fuel capacity was 14,320gal (114,620lb), used as the main heat sink in cooling the cockpit, avionics and other parts. The pilot was provided with a periscope, the indirect-vision screen being in the centre of the main panel. Behind him, likewise on upward-ejection seats, were the aft-facing navigator and reconnaissance/bombing systems operator.

The structure was almost entirely of brazed honeycomb sandwich, as used on the Type 720 interceptor, but in stainless steel. A vast amount of structural and systems research was completed in 1955–7, the periscope being tested on an Ashton, with Armstrong Whitworth making the Type 731 test vehicle and Bristol being assigned the engine/intake test vehicle Type 188, as already noted. In April 1957 the first 730 was in final assembly, for first flight in November 1959, when Defence Minister Duncan Sandys said no more fighters or bombers would

ever be needed, so the whole programme was cancelled.

Yet in that same year of 1957 studies began which were to lead to another supersonic reconnaissance-bomber for the RAF. This aircraft, called TSR.2 (from tactical strike and reconnaissance), was the subject of the most extraordinary display of government idiocy and hatred in history, and after virtually all the problems had been solved and the aircraft was in full production, half by Vickers (Supermarine) at Southampton and Weybridge and the other half by English Electric at factories near Preston, the whole programme was destroyed on 6 April 1965. Of course all the jigs and tooling had to be torn up, the aircraft already built or on the line were cut up for scrap, and even a request to keep one for a limited programme to test systems and equipment at over Mach 2 was refused.

The agonising politics of the TSR.2 programme are outlined in another PSL book, *Plane Speaking*, so here I will merely touch on the design features. TSR.2 did not have to meet such a severe range requirement as the Avro 730, but it was still formidable. Among other demands were the ability to hold Mach 2.2 for 45 minutes, to fly over 4,250 miles without air refuelling, and to achieve a mission radius, with a low-level supersonic dash, of 1,152 miles (1,000 nautical miles). The engines were two Bristol Siddeley Olympus 320 turbojets, each rated with afterburner at 30,610lb. This was high enough for STOL performance, despite the need for a small highly loaded wing for smooth flight at Mach 1 at sea level. The wing was almost a 60° delta, with downturned tips. Span was 37ft, area 700sq ft, t/c 3.7 per cent, and the structure had five spars and eight attachment links on top of the fuselage, all of them permitting limited movement. The challenging demand to operate from unpaved 3,000ft strips was met by extremely powerful blown flaps and the high thrust/weight ratio, maximum weight being 95,900lb. TSR.2 was really a Mach 3 aircraft, but normally held to Mach 2 because of its light-alloy structure. Other features included tailerons and a slab vertical tail, two tandem Mk 8 seats, internal and external weapon carriage, and the most comprehensive mission avionics ever devised up to that time. First flown on 27 September 1964, TSR.2 could carry bombs faster and perhaps lower than any other aircraft before or since, which means it had high 'penetrability'.

Having killed off our own bomber, the British government instantly replaced it with the General Dynamics F-111, while Dassault and Rolls-Royce tried

An artist's impression of the Avro 731, the test aircraft which was to explore many features of the Avro 730 bomber, including the basic configuration.

to sell a Spey-engined version of the Mirage IVA. This aircraft replaced the unsuitable *Vautour* (vulture) as the delivery system of France's *Force de Frappe* (deterrent force). The original requirement was for an aircraft to carry a nuclear bomb just over 17ft long to a target 1,242 miles away. Dassault had three bites at the cherry, first small (two early SNECMA Atar engines, 670sq ft tailless delta wing), then big (J75B engines, 1,595sq ft), and finally in between (15,430lb Atar 9K, 839sq ft). The original (small) Mirage IV prototype first flew on

17 June 1959, and by the end of the year had reached Mach 2 at 59,000ft, and had also made hard manoeuvres at 818mph at 10,000ft. Eventually two batches of Mirage IVA bombers were ordered, totalling 62 aircraft. Their deficiency in range against most potential targets was made up by flying in pairs, one refuelling the other. Bogie main gears enabled missions to be flown from dispersed strips compacted by quick-spray chemicals, the take-off run being shortened by batteries of 12 rockets. The aerodynamics and general technology rested on the Mirage III, but in many respects the Mirage IVA broke new ground, and over the years survivors were repeatedly upgraded and converted for fresh roles, notably as the Mirage IVP to launch the ASMP cruise missile.

In the Soviet Union V. M. Myasishchyev breathed a huge sigh of relief when he finished working on the

103-M (service designation M-4) subsonic heavy bomber, which was an attempt to meet a range requirement which in the early 1950s was impossible. He could not relax for long, because his next assignment was a supersonic heavy bomber. This aircraft, the M-50, started in 1955 mainly as a response to the B-58, posed numerous challenges, and required totally new techniques, equipment and even materials. The Central Aero-Hydrodynamics Institute assisted with the overall layout, and after studying 'more than 30 configurations, including exotic canard ones' it settled down as a conventional tailed delta, distinguished by the impressive 187ft length of the circular-section fuselage. After rejecting 12 engine arrangements, three were studied in detail. After considering pairs of engines above and below the wings, and two engines under the wings and two in the tail, it was decided that for minimum weight and drag the best answer was two under the wings and two on the broad tips. The specially developed P. F. Zubets engine was not ready, and Dobrynin's earlier VD-7 had to be used in the prototype when it first flew on 27 October 1959. The pilot and navigator sat in tandem seats which not only ejected downwards but were also the normal means of entry and egress.

The t/c ratio of the wing varied from 3.7 down to 3.5 per cent, and with an area of just over 3,000sq ft it was the largest supersonic wing built at that time. Fuel capacity was about 31,500gal, or about 250,000lb, out of a take-off weight of 441,000lb. Large tanks immediately behind the cockpit and in the tail were used to manage CG position and trim the aircraft throughout flight. By chance it was found that the landing gears could use major parts identical to those of the M-4, with tandem bogie main gears and small twin-wheel stabiliser gears near the wingtips, but it was modified in one major respect. To achieve the required angle of attack quickly and in a positive way on take-off, and thus minimise field length, the forward main leg was automatically extended by some 6ft upon reaching rotation speed. Myasishchyev called this 'the leaping bicycle'. Another claim to fame was that the M-50 was the first aircraft in the Soviet Union to be controlled by a fly-by-wire (FBW) system. In the prototype this system was backed up by complex and heavy mechanical flight controls, but, following extensive work with simulators, the production aircraft was to dispense with these. The system drove conventional ailerons and huge 'slab' tail surfaces. No particular problems were encountered, and pilots Goryainov and Lipko appeared to like almost

everything, but in the event it was decided not to put the M-50 into production.

Something similar happened in the USA, but in this case the aircraft was in a different class. When I first wrote about the North American XB-70 Valkyrie I declared, 'For the first time since the Wright brothers it was simultaneously the world's heaviest aircraft, the most powerful, the most costly and, except for the air-launched X-15, the fastest; it only just missed being the longest, and having the longest range.' It also had the biggest fuel system, the most powerful accessory systems, by far the biggest variable-geometry surfaces, and many other new features, all designed for operation at temperatures where lead would melt. One gets the impression the B-70 was a cut above the ordinary.

The programme began in the early 1950s, when the USAF Strategic Air Command (SAC) bore the West's sole responsibility for what was called 'deterrence'. Until the 1960s, its chief delivery vehicle would be the B-52, but that was considered to be increasingly deficient in speed and height, and was thought unlikely to serve beyond 1962 (little did they know!). The B-58 was deficient in payload and, especially, in range. So after prolonged studies two giant programmes were launched in February 1955. Weapon System WS-110A called for a chemically powered bomber (CPB) with at least the speed of the B-58 and the payload and range of the B-52. WS-125A called for a nuclear-powered bomber (NPB), with B-52 payload and the unlimited range and endurance conferred by nuclear propulsion, even if speed was subsonic. In fact, the NPB proposals were both subsonic and so do not concern us.

In November 1955 Boeing and North American Aviation were awarded Phase I contracts for the CPB. Both already knew that to fly from a standard SAC runway on a 6,910-mile (6,000 nautical mile) mission with a supersonic aircraft was beyond the mid-1950s state of the art. As with the B-58, it was necessary to cheat, but on a bigger scale. NAA's answer seemed to be to take-off as one aeroplane, and penetrate enemy airspace as another. At takeoff the huge canard aircraft would be accompanied by extra outer wings, each pivoted to the stubby centre section and carrying a fuel pod almost as big as a B-47 fuselage and weighing 190,000lb, the overall span being 260ft. When within 1,000 miles of the target the outer wings, with their empty fuel tanks, would be jettisoned, leaving a stubby-winged bomber tailored to Mach 2.3. The landing weight would be 220,000lb, less than 30 per

cent of the 750,000lb takeoff weight, so that the wing and canard would be adequate for a normal landing. But nobody was really enamoured of the proposal, and when it was shown to Curtis LeMay, famed commander of SAC, he removed his cigar and said: 'Hell, this isn't an airplane, it's a three-ship formation'.

There ensued some of the most widespread research in the history of aviation. It embraced Zip fuel, as mentioned in connection with the F-108 Rapier. General Electric designed versions of the J93 both for this fighter and for the CPB. A variable-stator engine appreciably larger than the J79, the J93 pioneered shaped-tube electrolytic machining (STEM) drilling for the air-cooling holes down the length of each high-pressure turbine blade. This engine was also intended to pioneer the use of Zip fuel, such as ethyl borane, whose greater energy promised not only higher speed but also greater range. When such bizarre fuels were tested, the difficulties were so intractable that they were soon confined to the afterburner only. Then, on 12 August 1959, the whole Zip programme was abruptly terminated. Both the Air Force and Navy had built enormous Zip production plants costing several hundred million dollars, which in the 1950s was a lot of money. In the end, the J93 was matched to a fairly conventional kerosene fuel called JP-6 which could soak for hours at 232°C (450°F) without turning into what are called cokes or sludges. This was crucial, because the fuel had to serve as the heat sink into which most of the unwanted heat taken from the avionics, accessory systems, refrigerated landing gears and cockpit had to be dumped.

What North American really wanted was some kind of major breakthrough, and they got it. One day in 1955 Alfred J. Eggers, of the NACA Langley Lab in Virginia, was mowing his lawn and pondering on L/D ratios at high supersonic Mach numbers. He hit on an idea he thought

The first TSR.2, with blown flaps part-extended and the nose gear on full left lock. *PRM Aviation*

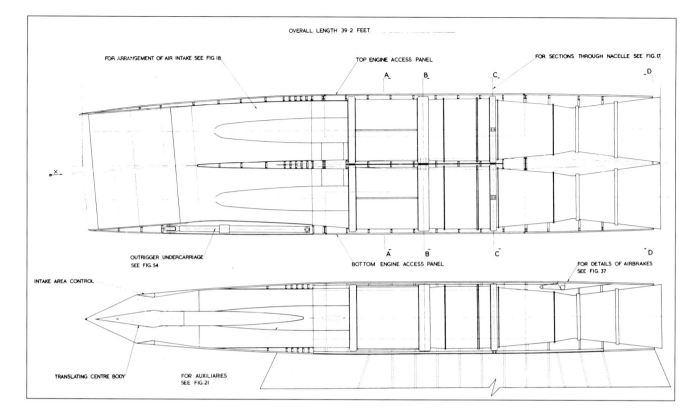

Side and plan views of the twin-P.159 nacelle carried on the wingtips of the original Avro 730. It shows such features as the variable inlets, outrigger gear and side airbrakes.

worth trying. The shape was that of a long, slender delta wing with an essentially flat top and with the fuselage *on the underside only*. There were two routes to higher L/D. One was to obtain the optimum interaction between the inclined shockwave from the leading edge and the pressure field under the wing. The other was to shape the fuselage to form a contracting duct, like a giant venturi, and Eggers saw that this effect could be enhanced if the outer wings could be hinged down to box the airflow in. Eggers and Clarence A. Syvertson jointly authored a paper in March 1956 entitled *Aircraft Configurations Developing High L/D Ratios at Supersonic Speeds*. Like the area rule paper of 1953, it was kept classified, because it was a major advance in aircraft design. Eggers called it 'compression lift', and it was precisely what NAA were looking for. With their very

first compression-lift tunnel model they got an L/D at Mach 3 just double the previous best at small angles, and a peak L/D 22 per cent higher yet reached at a smaller AOA. The result was that, instead of being unable to fly the mission even while staying subsonic for 90 per cent of the time, the bomber could now cruise at Mach 3 all the way.

From here it was mainly a matter of designing the hardware, and it was the biggest engineering task in the history of aviation. The only major flaw in the programme was that by the late 1950s it was self-evident that mere speed and height would very soon not confer immunity against anti-aircraft missiles. Amazingly, not only was this ignored with the B-70, but it was again ignored with the B-1 which followed it (Chapter 10). Instead, the heated political debate revolved around whether the emergence of the ICBM had made the concept of the strategic bomber obsolete.

For a comprehensive description of the XB-70 the reader is referred to *Flight International* for 25 June and 2 July 1964 (the author, writing under a pseudonym, was J. Philip Geddes). The vast wing, of 6,297sq ft area, was an almost pure 65.5° delta. Underneath was the colossal engine box, 7ft deep, 37ft wide and 110ft long. At the front was the biggest arrangement of variable inlets ever built, from which the ducts curved inwards to leave cooled compartments outboard for the bogie main gears. The fuselage was almost entirely ahead of the wing. The four-man crew in the nose each had an individual escape capsule which could be sealed before ejection. Behind were the main avionics bays, aft of which were huge tanks of water and ammonia to help the fuel to absorb unwanted heat extracted by the environmental system. Flight controls comprised four inboard and two outboard elevons on each wing, a large trimming canard foreplane with trailing-edge flaps, and twin powered vertical tails. About 70 per cent of the structure was stainless steel, with a surface like a mirror. There were over 600 services driven electrically, all capable of operating at 332°C (630°F). A new high-temperature hydraulic fluid, Oronite 70, served 85 linear actuators and 44 hydraulic motors, the 12 most powerful of which drove the outer wings to anhedral angles of 25° for transonic acceleration, and 65° for Mach 3 cruise. The bomb bay had a rearward-sliding door and could carry two B53 high-yield bombs, or various other weapons.

GE got the first J93 engine on test in September 1958. Weighing 4,770lb, it was rated at 27,200lb thrust with afterburner, running on JP-6 fuel only. In 1959 the WS-110A programme was 'reoriented' to a single XB-70 prototype, though later a second was added. For a while an RS-70 reconnaissance-strike aircraft was the subject of further arguments, but this was killed by ICBMs and the SR-71A. The six engines for the first XB-70 were installed from June 1962, but so many were the problems that the first flight was delayed until 21 September 1964. On that occasion the main gears refused to retract, one engine showed itself as overspeeding, and was shut down, and on landing the anti-skid system malfunctioned and blew two main tyres, but all this gives a false impression. The flight test programme was really impressively successful. Col Joe Cotton, the Air Force project pilot, likened it to 'driving a Greyhound bus 200mph around the track at Indianapolis'. On its 39th flight in 1966 the No. 2 aircraft held Mach 3.05 to 3.08 for just short of 3½ hours. Tragically, this aircraft was lost on 8 June 1966 from a mid-air collision with an F-104, which approached too close and became caught in the powerful pressure field round the outer wing. After gathering truckloads of data for the American SST programme, the No. 1 aircraft flew to the Air Force Museum on 4 February 1969.

There followed many years of study. One wag said that AMSA, the Advanced Manned Strategic Aircraft, really stood for 'America's most studied airplane' – which eventually led to the B-1. Even then the study was not deep enough, because this was planned for Mach 2 at 50,000ft and later had to be redesigned into the B-1B for 600mph at a height of 200 feet (Chapter 10).

The only other country with an interest in SSBs remained the Soviet Union. It was probably correct not to put the Tu-98 into production, because in the 1950s the only likely enemy was the United States, and this quite modest aircraft could not have flown strategic missions. (It could, however, have operated against Turkey, Iraq, Iran, Afghanistan, Pakistan, India, South-East Asia, China and even Japan.) Instead, the Tupolev bureau was instructed to design a much larger aircraft with a mission radius as near to 3,000km (1,864 miles) as possible. Work began immediately upon cancellation of the Tu-98 in 1955. At the 1961 Aviation Day ten of the Tu-105 prototypes flew over, one carrying a recessed supersonic cruise missile. It was assessed by Washington as a strangely useless aircraft, because with an estimated range of 2,500km (i.e. a combat radius of well under 750 miles) it could not reach any potential target. It took 25 years for Western analysts to comprehend that, despite having only two engines, this aircraft, designated Tu-22 in air force service and dubbed 'Blinder' by NATO, was actually considerably bigger and heavier than the B-58.

Made almost entirely of conventional aluminium alloys, the Tu-22 was given a giant area-ruled fuselage over 136ft long, with a circular section almost entirely filled with fuel, the normal fuel load being 94,578lb. The conventional swept wing was mounted quite low, there being just sufficient room under it for the bomb bay, or for recessed cruise missiles. Surprisingly for a supersonic aircraft, the bogie main gears were arranged in typical Tupolev style to retract backwards into large fairings aft of the wing. Control surfaces comprised ailerons, one-piece tailplanes used for pitch only, and a rudder. The unique feature of the design was that the large afterburning turbojets were placed on each side of the vertical tail, above the rear fuselage. Drag would have been reduced by putting them inside the fuselage, but mounting them above saved the weight of a huge inlet/duct system. No attempt was made to fit variable supersonic inlets; they were made plain holes, but with the front ring able to translate (move) forward before take-off to admit extra air around the sides. The crew of three were arranged in

Even today the XB-70 is perhaps the most impressive aeroplane in the world, and it is sad that it never thrilled airshow crowds. *PRM Aviation*

The pioneer adjustable-sweep jet aircraft was the Messerschmitt P.1101, which was nearing completion at the company's Oberammergau research centre when Germany was overrun in 1945. Cutting a long story short, this was the basis for the Bell X-5, which first flew on 20 June 1951. Whereas the P.1101 had wings adjustable on the ground only, from 35° to 40° and then to 45°, the Bell X-5 could sweep its wings in flight, and from 20° to 60°. Each wing was pivoted to a central frame and connected via an irreversible screwjack to a central gearbox. This mechanism was ahead of the pivots, but in some much later VG wings the drive screwjacks were at the rear. In the X-5 the gearbox also moved the wings bodily to front and rear, so that at 60° the roots had moved forward about 27in. The X-5s were successful, but subsonic, and lethal in a spin.

British work, mainly by B. N. (later Sir Barnes) Wallis, at Vickers-Armstrongs, began as early as 1945. His first major VG programme was Wild Goose, followed by a more ambitious Swallow project for a 'paper dart' supersonic transport or bomber with engines pivoted near the wingtips. An offshoot was the Vickers submission to Operational Requirement 346, with VG wings, four RB.153 engines in a square group at the tail, and a 'droop snoot' hinged nose. The main feature of all these VG aircraft was that at last the wings were merely pivoted, with no need for the pivots to slide to front or rear. Unfortunately for Britain the government, having announced in 1957 that military aircraft would no longer be needed, instructed Vickers to send all their VG data to NASA, the newly formed US National Aeronautics and Space Administration. This was done in December 1958. This information was of considerable help in Langley Research Center's subsequent work on VG wings, which resulted in the F-111 (Chapter 10), though it is not mentioned in NASA's own account of *Fifty Years of Aeronautical Research* (NASA, 1967).

By 1960 the general geometry of such wings was becoming common knowledge, as were the enormous benefits. Studies in the Soviet Union in 1962 showed

One of the original Tu-22 supersonic bombers, equipped for free-fall bombing only. *PRM Aviation*

that several existing military and even civil aircraft could be transformed by fitting VG wings, even though the pivoted portions would have to be at the ends of a very large fixed centre section. In the end two aircraft were picked for such modification, while other teams designed VG aircraft from scratch. These modified IG (*Izmenyaemya Gayometriya* = VG) aircraft were based on the Su-7 and Tu-22. While the Su-7 derivatives quickly produced results, the Tu-22 studies, in particular the Tu-106 with NK-144 engines, eventually made it clear that for the best results a completely fresh design was required. This went ahead in 1967 under the deliberately misleading designation of Tu-22M, and the first edition of this book described the Tu-22M as a VG conversion of the Tu-22.

The supposed VG conversion of a Tu-22 was seen by a reconnaissance satellite in 1969, outside the gigantic factory at Kazan, and given the NATO name Backfire. Although designated as the Tu-22M-0, this was actually a completely new aircraft. The first of nine began its flight test programme on 30 August 1969. In round figures, even the Tu-22M-0 could carry double the weapon load of the Tu-22, about 40 per cent further, with a high-altitude dash capability of Mach 1.69 in early versions, and nudging Mach 2 in the final variants. The outer wings were on pivots about 35ft apart, at 50 per cent chord, with fences outboard of the pivots near the outer edges of the fixed inboard wing. Each outer panel could

be swept to 20°, 30°, 50° or 60°, and had a full-span slat, aileron, spoilers/lift dumpers and high-lift flaps. The Tu-22M-0, M-1 and M-2 were powered by twin Kuznetsov NK-22 afterburning turbojets, each rated at 48,500lb. These were side-by-side in the rear fuselage, fed via giant ducts passing over the wing leading from fully variable inlets of the upright type, with an inner splitter plate to separate the boundary layer and form an inclined shock

A Tu-22M-3, with pylons empty and wings at maximum sweep. Loss of one in Georgia in August 2008 highlights the need for an update. *PRM Aviation*

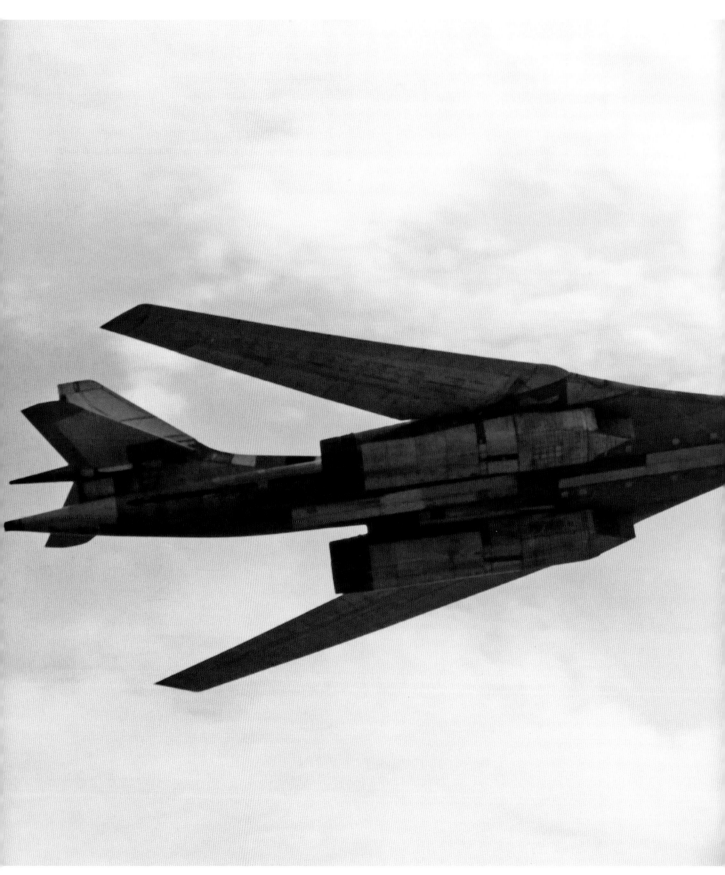

The Tupolev Tu-160 'Blackjack'. *PRM Aviation*

focused on the outer lip. The main landing gears each had six wheels with high-pressure tyres, and retracted inwards into the fuselage under the engine ducts. Ahead of the fuselage tanks, avionics bays and environmental system was the cockpit for a crew of four, with 'gull wing' upward-opening doors. In front was the pointed radome and flight-refuelling probe.

By 1972 the first major production version, the Tu-22M-2, was in flight test. This introduced more efficient engine inlets, with upper and lower boundary-layer ejectors, and totally revised avionics. The main production version, the Tu-22M-3, introduced the NK-25 engine, rated with maximum afterburner at 55,115lb. All versions had an internal weapon bay, as well as the ability (if necessary by fitting different bay doors) to carry a cruise missile recessed under the fuselage and two further similar missiles on pylons under the extremities of the fixed wing. If necessary, racks could be added under the fuselage on each side of the weapon bay for 12 bombs of 1,102 1b each, or other loads. Defensive tail armament was retained: the Tu-22M-2 had two twin-barrel GSh-23 guns, while the M-3 version had a single GSh-23 mounted on its side (i.e. with barrels superimposed), in a low-drag installation.

By 1992 about 400 of all versions had been delivered, of which about 100 had been kept serviceable in the VVS and AV-MF (naval aviation). Ukraine kept a further 20. The larger Tu-160 'Blackjack', and the USAF's B-1B, are discussed in Chapter 10.

Chapter 9

SSTS

If designing an SSB is hard, designing an SST (supersonic transport) is even harder. To the demands of supersonic cruising speed over a long range, and the ability to fit into established traffic patterns and use existing runways, are added comfortable accommodation for passengers of any age or fitness, virtually 100 per cent safety and reliability, and operating costs sufficiently low to make the aircraft saleable. But it is much harder even than that. Indeed, until recently many would claim that a satisfactory SST was incapable of being built. Ignoring claims of environmental damage which are at least not proven, and are probably nonsense, the noise of an SST near an airport tends to be significantly worse than that of other aircraft, and certainly above the limits set by certification authorities. In addition, the intractable problem of the sonic boom has resulted in a general agreement not to fly at supersonic speed over land, severely limiting the routes over which SSTs can realise their intended potential. Not least, the soaring price of crude oil since 1973, and more dramatically today, has hit the SST harder than any other transport aircraft.

Hope springs eternal. Whereas over 30 years ago the concept of the SST was extremely unpopular around the world, except to an army of knockers and denigrators to whom it meant publicity and sometimes an attractive income, today the onward march of technology has made possible SSTs with dramatically improved economics and better environmental acceptability. The likelihood of such aircraft was enhanced by the 30 years of uneventful operation of Concordes, which demonstrated that most of the objections of the opponents were without foundation. It was unfortunate that this pioneer SST should have been withdrawn from use prematurely because of a disaster which had nothing to do with the fact that it was an SST.

Each chapter of this book naturally treats its subject in chronological order, but even with something as challenging as an SST several things have been going on around the world simultaneously since the mid-1950s. The first large formalised study was carried out in Britain. In 1956 the Supersonic Transport Aircraft Committee (STAC) was set up by the Royal Aircraft Establishment (RAE) at Farnborough. Its chairman was the RAE's Deputy Director, M.B. (later Sir Morien) Morgan, and its members included nominees from manufacturers, airlines and ministries. A dozen companies joined the RAE in carrying out research into aerodynamics, structures, materials, propulsion and many other topics. In March 1959 the Committee reported to the Minister of Aviation, recommending the construction of two types of SST. One was a 100-seater to cruise at Mach 1.2 (800mph) for 1,500 miles. It had a zigzag wing, understandably called an M-wing, the engines being grouped at the forward-projecting kinks. Its main advantages were fast and relatively cheap development, and the ability to cruise 50 per cent faster than existing jets without causing sonic booms heard on the ground. The other proposal was for a slender delta, with engines grouped at the back, to carry 150 passengers at Mach 1.8 (1,200mph) for 3,500 miles.

The Ministry awarded modest study contracts, but 1959–60 was a time of politically induced turmoil in the British aerospace industry, for almost every firm was either trying to merge with others, or trying to avoid doing so. Nevertheless, a little tunnel work was possible and, while the M-wing was quickly dropped, the Mach 1.8 SST looked increasingly attractive. From the start it had been envisaged as a delta of exceedingly low aspect ratio. This book has emphasised that there have always been plenty of aerodynamicists eager to predict disaster, and the HP.115 research aircraft was built to placate

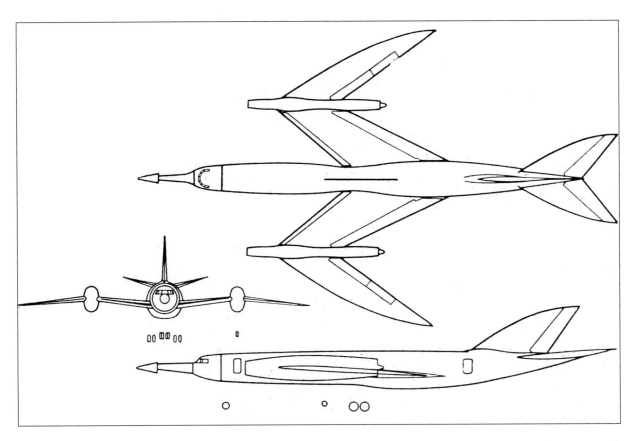

them, as related later. However work at Farnborough and in industry showed the excellent properties of a refined form of slender delta, which began as the Gothic delta, the shape of several Bristol 198 SST studies going back to 1959. This shape, similar to a Gothic cathedral window, was then developed into the ogee, or ogival, delta, with a graceful leading edge, which starts with extreme sweep, curves out and then curves back. One Farnborough researcher, W. E. Gray, explored high-AOA behaviour with 'paper dart' models thrown by hand!

One of the difficulties of the SST is that, like other transport aircraft, the wing has to be designed for cruising flight. With subsonic aircraft the wing can also be matched to take-off and landing, by adding reasonably light and reliable high-lift devices, but with the SST wing this is impossible unless the wing has variable sweep. Vickers-Armstrong were world leaders in VG wings, but they had no major role in British SST research after delivery of the 1959 Report. The study contracts were placed with Hawker Siddeley, a previously existing group, and with the newly formed British Aircraft Corporation (BAC) which embraced the firm that had pioneered all the slender-delta work,

In the author's view the British so-called M-wing SST for low supersonic speeds was never a viable proposition. The nose spike made a shock cone ahead of the outer wings.

Bristol Aircraft. Working with Farnborough, Bristol began with Gothic Type 198s, often with canards, with six or eight Olympus engines grouped above the wing. The study moved on to an ogee 198 with six Olympus in two widely separated groups below the wing, and then, in January 1961, scaled this down to the Bristol 223 with four Olympus. It is interesting to note that the 198 was cut down in size because of doubts about the economics of six engines and the boom intensity of a 385,000lb aircraft. The irony is that the eventual Concorde, derived from the Bristol 223 with only four Olympus, had the same span as the 198, greater performance, and a take-off weight of 409,000lb.

Before continuing the history it is worth looking further at why such a wing makes sense. There are numerous interacting factors, so that it is difficult to know where

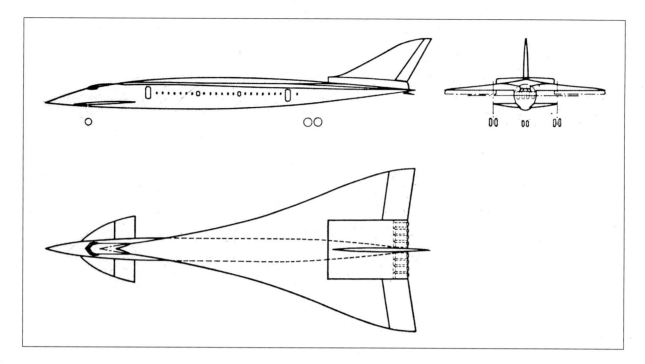

One of the earliest British Mach 2 SST projects, with some good features and more bad ones.

to start. We can be certain that, for minimum wave drag at supersonic speed, the thickness and the lift of the wing should be distributed along the greatest possible length from front to rear, but nobody would want a rectangular wing with, for example, a chord of 150ft and a span of 30ft! The slender delta is aerodynamically much more efficient, and its leading edge lies behind the Mach cone generated in cruising flight. On the other hand, at low speeds, at a large AOA, such a wing would appear to be almost useless, resulting in unacceptably high take-off and landing speeds, and the need greatly to extend the world's runways. The breakthrough was the discovery of how vortex lift can be put to use, two of those responsible being D Küchemann, whom we met in Chapter 5, and E. C. Maskell. At very low AOA, as in cruising flight, the air flows across a slender delta wing normally, without flow separation. As AOA is increased, at lower speeds, the air flowing up and over the leading edge swings out as it climbs on top of the wing, then swings inwards and finally outwards

again. This curved path is most pronounced near the wing, and is greatest of all in the boundary layer. Eventually, the streamlines in the boundary layer, encountering rising pressure to the rear as well as inwards, are forced to curve outwards along a separation line. Along this line the airflow no longer follows the wing surface, but rises off it in a vortex sheet. This writhes back across the front part of the wing, starting at the tips, and eventually progressing all the way in to the root. On a humid day the vortices are stridently visible as white condensation clouds. Concorde passengers saw them suddenly blotting out the view on take-off rotation.

Normally such flow separation and vorticity would be highly undesirable, but for our slender delta it can be put to good use. This is because, being attached to a sharp leading edge, the separated flow and subsequent vortex sheet are stable, steady and controllable, remaining qualitatively unchanged throughout the entire speed range, and changing quantitatively only with aircraft speed and AOA. Thus, while being non-existent (or existing in an embryonic form only at the very tips of the wing) in Mach 2 cruise, the huge and powerful vortex considerably increases suction (lift) over the wing upper surface at low speeds. Without the vortex, a supersonic slender delta wing might generate an L/D ratio of 4 on landing. The vortex raises this to between 6.5 and 7.2, using 1960s Concorde technology, and the best 2008 figures are around 9. Bearing in mind the difficulty of fitting slats or droops to a slender delta, it is fair to say that an SST of this type could not fly without its vortex.

Nobody gets much for nothing in aviation. Whereas other factors call for s/l (aspect ratio) to be as low as possible, drag due to vortex formation is reduced by increasing aspect ratio. An accompanying figure shows the four sources of drag in Mach 2 cruise for wings of differing aspect ratio. The wing with βs/l=0.8 would look rather like that of a Sukhoi Su-27. At the other end βs/l=0 means that span is zero, where (apart from lift also tapering off to zero) vortex drag is infinite. Clearly the best compromise is somewhere in between. Friction drag stays sensibly constant, varying only with wetted area (surface scrubbed by the boundary layer). Wave drag due to thickness, or the volume of the wing, becomes unacceptable with large spans, and cannot reasonably be kept within bounds by making the wing as thin as cardboard. Wave drag due to lift is acceptable and inevitable. The problem is that, as aspect ratio is reduced, vortex drag due to lift rises, reaching unacceptable levels at below βs/l=0.2. For lowest overall drag in Mach 2 cruise, βs/l should be around 0.4, and if β is 2 (a typical value) then the semi-span comes out to about one-fifth of the root chord, which is typical of SST wings.

One can express the same ratio by saying that the root chord is five times the semi-span. This is just the opposite of the trend with subsonic jetliners, whose root chord is about one-third of the semi-span. Obviously, with such an enormous root chord, not much fuselage will project in front or behind. The view for most passengers will be nothing but wing. There is no point in adding a horizontal tail. Instead, the trailing edge is placed across the airflow and fitted with elevons, and this also fits in with the need to shed the vortex sheet cleanly off the trailing edge at low speeds. A side issue is that, whereas conventional wings 'run out of steam' and stall at around 18° AOA, the slender delta shows no such tendency. The vortices even increase the lift-curve slope to beyond 25°, and the wing is still lifting strongly at 60°. Unfortunately, this marvellous extra lift has so far proved to be out of reach. With a root chord covering most of the length of the fuselage there is no evident way of pivoting the wing so that its incidence can be varied, and there is no way an SST can rotate its fuselage and passengers to 60° AOA and then touch down!

Above: Simple representation of the vortex flow past a slender wing lifting at high AOA. The vortices roll up actually <u>over</u> the wing and dominate the flow pattern.

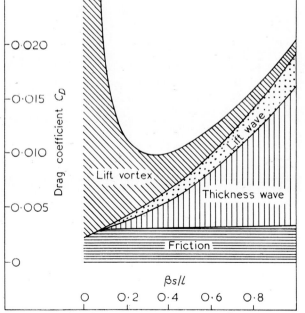

Left: Sources of drag at Mach 2. The aircraft on the right would have wings of long span, while the left margin (βs/l=0) means span is zero. An SST would naturally be designed near the lowest point on the total drag curve.

Back in 1960 the ogee wing planform appeared to be a natural candidate for an all-wing aircraft. Accordingly, the Hawker Siddeley study investigated 'integrated' designs in which the wing was thickened inboard and blended into a pressurised (inhabited) area which projected slightly at front and rear to carry the cockpit and vertical tail. Such a scheme might have worked in a giant SST over 400ft long, but in the size being considered it was a non-starter. This meant that there had to be a wing of minimum volume and a conventional fuselage, which also had the advantage of enabling the passengers to have windows. As the presence of the fuselage generally causes a suction field on the wing, it was logical to place the fuselage above the wing. This gives a useful increment of lift without

increase in drag, though, as noted above, it spoils passenger view.

Further benefits were gained from the engine installations. Perhaps for minimum drag the integrated answer was best, because the engines could be inside the aircraft, but external nacelles can give advantages. At Mach 2 the inlet throat is smaller in cross-section than the compressor inlet or nozzle exit area. Inside the inlet the shockwaves slow the flow to subsonic speed, so the expanding duct increases the pressure. Outside the flow remains supersonic, and the acceleration of the flow around the nacelle causes a high-pressure field. With care it is possible to reduce drag by reduced pressure over the front of the nacelle, while increasing lift by placing the nacelle, with its main high-pressure field, under the wing. To achieve the desired smoothly expanding tube shape it is necessary to bury the top of the engine in the underside of the wing; this is no problem (diagram, Chapter 8). All engines could be grouped near the centreline, but splitting them into two nacelles well out under the wing makes it easier to achieve the optimum nacelle shape, avoids ingesting material thrown up by

Hawker Siddeley worked on SST designs with a fuselage merged into the wing. This particular project had twenty-four engines!

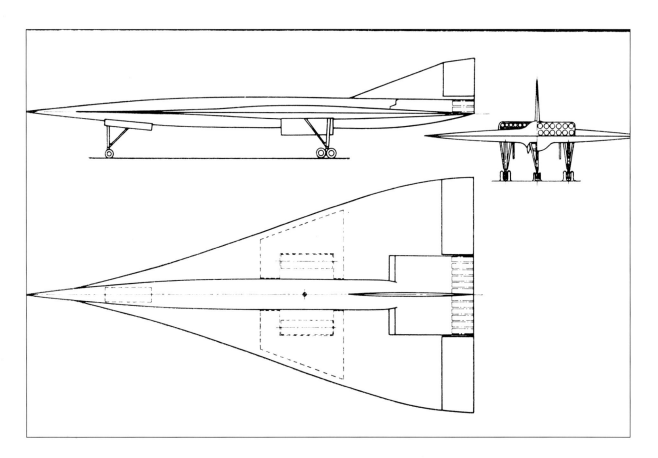

Pre-production Concorde 01 being readied at Filton.
An amazing achievement in its day, it has since been
overtaken by a vast amount of later knowledge. *Airbus
UK/PRM Aviation*

the nose gear, and reduces the thickness of the boundary
layer that must be prevented from entering the inlet.

When in October 1960 the newly formed BAC received
their SST study contract – for £350,000, whereas today
£350 million would go nowhere – they were instructed
to seek foreign partners. The general opinion in the
USA was that the correct cruise Mach number was 3,
as will be related later. Federal Germany concluded
that its industry was not ready for such a challenge. In
France, however, not only was there already an SST
project, but it was almost a mirror image of the Bristol
223. Sud-Aviation, proud builders of the Caravelle, had
in 1957 received an Air France outline requirement for
what was naturally called the Super Caravelle, carrying
the same 80-odd passengers for up to 2,000 miles. At the

1961 Paris airshow virtually every visitor was fascinated to see, high above the company stand, a model of this aircraft. Bearing in mind that not a word or artist's impression of the enormous British work had leaked out, this was the first SST image – or at least the first ogival delta – the world had seen. Sud announced that they were working with Dassault, and believed that the Super Caravelle could be 'ready for service by 1964–5'. This was optimistic, by 13 years.

In general I want in this book to give readers a flavour for the subject of supersonic flight, and at least a basic understanding of the technicalities. From this point on we enter the programme which led to Concorde. My friend Capt Edouard Chemel, former Chief Pilot of Air France, has a large house completely filled with many thousands of books and tracts about Concorde, in almost every language. This makes me feel there is little point in going into detail on it. But it is a great story of dogged determination to win, not only over the technical problems, but also over a strident army of doubters and denigrators. As Concorde was not American, it received virtually no considered or objective publicity around the world, which one might have thought a comfortable public transport vehicle travelling at twice the speed of sound merited. The publicity it did get was almost totally adverse, especially in Britain. In a review of the past 50 years Jack Steiner, one of the greatest of Boeing's former leaders said, 'This very large and long-term achievement has not been given the credit in this country [the USA] that it deserves. The Concorde accomplishment, I feel, is a very substantial one.'

In 1961 BAC urged Sud-Aviation to agree on a common SST, while Sud-Aviation said: 'Certainement, mais il y a deux avions.' They insisted on a 2,000-mile SST (at the time called medium-haul, but today regarded as short-range), perhaps tackling longer ranges later with the benefit of experience. BAC was convinced that a viable SST had to have transatlantic capability. There would have been an impasse, had it not been for the crucial fact that the engineers kept talking. Led by Pierre Satre and Lucien Servanty on one side and Doc (later

One of the Air France Concorde fleet, in take off or flyby configuration. The leading edge is fixed, and is creating a visible vortex. *Air France/PRM Aviation*

Production Concordes featured a hinged nose and sliding windscreen visor to improve visibility from the flightdeck during the high AOA landings. *Air France/PRM Aviation*

Sir Archibald) Russell and Bill Strang on the other, their frank talks opened their minds to the viewpoint of their partners. At last Strang and Servanty were figuratively locked in a room at 37 Boulevard Montmorency in Paris for a whole day. When they emerged, they brought with them the first general-arrangement drawing of Concorde. So on 29 November 1962 an agreement was signed between the governments of France and the UK 'for the development and production of a civil SST aircraft'.

The agreement was split 60/40 in favour of France, reflecting the British dominance (once suggested at 70/30, but eventually 60/40) on the engine. There was never much argument over the latter; the Olympus turbojet, already developed in what appeared to be a suitable thrust category for the Mach 2.7 TSR.2, had

even been picked for the Super Caravelle. But there were still two Concordes within the same outline. Sud Aviation persisted with the short-hauler (100 passengers, 220,000lb) while BAC worked on the transatlantic aircraft (90 seats, 262,000lb). Feedback from airlines at last convinced the French that there was simply no market for the short-hauler, and indeed even the long-haul version had to be repeatedly upgraded. In May 1964 work went ahead on a common version weighing 326,000lb, with engines redesigned for much greater mass flow, giving 32,400lb thrust with modest afterburner. This was approximately the standard of prototypes 001, flown at Toulouse on 2 March 1969, and 002, flown from Filton to the British test base at Fairford on 2 April 1969. Subsequently the aircraft grew further, the most visible changes being to the cockpit windows and the addition of a 13ft tailcone, while weight grew to 409,000lb. This was made possible by a series of changes to the engine.

As finally developed, the Olympus Mk 621, which logged many times more Mach 2 experience than all the world's other engines combined, had a mass flow of 410lb/s, compared with 181 for the first engine proposed for Concorde. Thrust is 39,940lb, and weight

5,793lb. Among major features are improved aircooled turbine blades, an annular combustion chamber which eliminated visible smoke, and a new afterburner (part of Snecma's French 40 per cent contribution), downstream of which is a noise-attenuating variable nozzle and upper and lower reverser buckets, which squeeze the jet on take-off, open it out in a supersonic diverging nozzle in cruise, and reverse the thrust after landing.

Features of the production Concorde included an improved 'droop snoot' hinged nose, integral tankage for 26,350 gallons of regular airline fuel, including trim tanks at front and rear to give dragless correction of the shift in aerodynamic CP twice on each flight, totally electronic control of the engines and their complex inlet and nozzle systems (the first in the world in regular service), a monofuel emergency power unit in the tailcone (replacing twin gas-turbine starters in the prototypes), and an interior configured for a flight crew of three (one man facing a right-hand wall panel) and 100 passengers. Whereas at the start of the programme many airlines rushed to reserve supposed production slots, the implacable opposition of the American industry, and in particular the Port of New York Authority, which refused to accept any Concorde service, ensured that no aircraft were sold except to the original two national airlines. Two

interim development aircraft were flown in 1971–3, and 16 production aircraft in 1973–9, each priced at (1977) £23 million. The first two production aircraft were never brought up to final standard, and were not delivered. I was fortunate enough to make six flights in the second of these 'not quite' production aircraft, G-BBDG, and liked the comment of the BAC navigator who said, at 59,000ft near Tehran, 'It's like a clockwork mouse; you wind it up and it goes!'

Despite all that the Port of New York Authority and other opponents could do, Concorde entered passenger service in 1976. The SST was still something calculated to get barricades put up in the streets. I was honoured to write the biography of that great engine man Sir Stanley Hooker, and he went to Bahrein on Concorde's first scheduled flight. He recalled 'The Minister of Foreign Trade . . . said to me, "What a great honour you have

The first Tu-144 was completely redesigned, resulting in a larger and heavier aircraft with engine nacelles separated and housing the (different) main gears. Here the added canards are extended. *PRM Aviation*

77115 was one of the last batch of Tu-144Ds, with much more efficient engines. *PRM Aviation*

done our country by bringing your magnificent Concorde here on its inaugural flight!" Considering the violent hullabaloo that was going on in New York, and that nobody else would allow us in, it was difficult to find the right answer!'

Very gradually the protesters found other things to hate, and the Concordes faded from the headlines. As well as halving the flight time on every route they flew, they became important vehicles for charterers, and many people flew at Mach 2 for sheer pleasure, obviously unaware of the terrifying list of dangers predicted by the antis. The total of Concorde passengers neared 5 million. Total flight time reached 243,845 hours, of

which over 192,000 were faster than sound. About 90 per cent of this total was at Mach 2, so Concorde logged far more flight time at 1,350mph than all the thousands of Mach 2 fighters and bombers ever built. Reliability was outstanding, and it was a strange irony of fate that the entire operation should have eventually have been terminated by damage caused by a piece of metal falling off a subsonic aircraft.

Indeed, for 24 years this extremely challenging vehicle was the safest aircraft in the world, with a perfect record. Then, on take-off from Paris CDG on 25 July 2000, an Air France Concorde with every seat filled ran over a large strip of titanium, part of a thrust-reverser on a DC-10 of Continental. It burst a tyre, which via a rapid and extraordinary sequence burst a fuel tank, and ignited the escaping fuel. Trying to reach Le Bourget, the aircraft crashed at Gonesse, close to where a Tu-144 had crashed in 1973. After various modifications, services were resumed on 7 November 2001, but by this time many factors, including the '9-11' terrorist attacks, had severely hit bookings. Air France flew their last SST service on 27 June 2003, British Airways following on 24 October. The last retirement flight was on 26 November 2003. All Concordes found homes, mainly in museums. Suddenly, man's fastest transport vehicle was 60 per cent slower.

Years earlier, the crash of a Tu-144 at Goussainville had played a part in ending the career of the only other SST to enter service. From the start, this was a purely Soviet development, begun in 1963, partly as a result of the Concorde programme, but also because of the size of the Soviet Union. It was calculated that, if Aeroflot had 75 SSTs, cruising at up to 3,000km/h (1,864mph), these could increase the average time saved (compared with surface travel) per passenger-journey from 24.9 hours to 36.6. Having established the need, a full-scale programme followed immediately, the design contract being placed with the OKB (experimental construction bureau) of A. N. Tupolev. At a very early stage the design targets were agreed as 121 passengers, to be carried 6,500km (4,040 miles) at Mach 2.35 (2,500km/h, 1,553mph). This was considered to be the limit for light alloys. At high Arctic latitudes the cruising speed was to be restricted to Mach 2.05, because of the warmer stratosphere.

A model at the Paris Salon in 1965 showed a tailless delta with a plan shape similar to that of Concorde, but rather simpler, with the leading-edge shape made up of large curves linked by two essentially straight lines, almost without camber. The four Kuznetsov NK-144 augmented turbofans, each rated at 38,580lb, were grouped in a single large box under the wing, with a deep gash at the front to accommodate the retracted nose gear. I questioned Andrei Tupolev on this point, and he assured me that they had considered the Concorde layout and that the single box gave minimum drag. The main landing gears were attached outboard of the engine box, each 12-wheel bogie retracting forwards to lie inverted in the extremely thin wing. Tupolev later said, 'We thought in terms of 2.5 per cent, but the Tu-144 prototype did not achieve this target.' Fuel capacity was 19,247 gallons, with front and rear trim tanks. Flight controls comprised eight elevons and two rudders, all approximately square and each driven by three power units. The opaque nose drooped 12° for take-off and landing. All wheels were braked, but no reversers were fitted although, as in prototype Concordes, there was a tail parachute.

The first prototype, 68001, was the first SST to fly, on 31 December 1968. It was accompanied by the 211, also called the A-144, a MiG-21MF rebuilt to test a model of the Tu-144 wing and elevons. Mach 1 was exceeded on 5 June 1969 and Mach 2 on 26 May 1970, the highest figure being 2.4 (never reached by any other SST). At least 15 Western companies supplied hardware for the Tu-144 systems, and in addition a fair amount of industrial spying certainly went on to bolster the Soviet storehouse of knowledge. An American author, Howard Moon, wrote a book on this topic, *Soviet SST*, (Orion Books, 1990). The reason for the espionage was Soviet uncertainty; the only certainty was that some design choices were mistakes. So the next Tu-144, 77101, did not appear until spring 1973. Western observers were intrigued to see that it was a completely different design, with only a general similarity to its predecessor.

The wing was no longer an ogive, but a double delta, with straight leading edges swept at 76° and 57°. On the other hand, where the first wing was sensibly flat, the new wing featured impressive camber and tip droop, the tips being broad and square-cut instead of Küchemann type. The t/c ratio was reduced, and the trailing edge cambered downwards, so that the fairings for the 16 power units were more prominent. The fuselage was completely redesigned, with a revised cross-section and structure, different materials, a new environmental system, and length increased by 20ft 8in, with 34 windows on each side instead of 25. The droop nose was completely new, with a modified profile, different drive system and better view in the cruise position. The engines were completely divided into two boxes, but because of the great length of the NK-144 the pairs of engines could not be far apart. Indeed, the separation was possible only because the inner wing was reduced in leading-edge sweep. The result was to separate the Nos 2 and 3 afterburner nozzles by the width of the fuselage. One result was that, everything else having been redesigned, it was no very great extra task to redesign the main landing gears. These became quite different bogies with eight larger wheels, pivoted to forged frames in the engine nacelles. Each unit first rotated 90° about an axis parallel to the fuselage so that, as the leg was pulled forward, the bogie lay upright in a narrow refrigerated compartment between the inlet ducts. The nose gear was moved forward no less than 31ft 6in, and lengthened to retract forwards into a long unpressurised bay under the forward fuselage. A major aerodynamic improvement was the addition of retractable canard foreplanes, normally recessed into the top of the fuselage aft of the flight deck. For take-off and landing, the canards were arranged to swing out to zero sweep, with sharp anhedral. Their span was then 20ft, and each surface was of high aspect ratio, almost untapered, with a double leading-edge slat and double-slotted flaps. They greatly improved lift and handling at low speeds, and allowed the elevons to give lift instead of a large downwards force. Fuel capacity was increased to 26,121 gallons, and the fore/aft trim-tank system was properly engineered, instead of being an afterthought.

Few aircraft have ever been so completely redesigned while retaining the same designation. MTO weight jumped by no less than 50 tonnes, to 396,830lb. Effective operating range, which in the prototype had been about half that specified, did reach close to the desired value with 140 passengers, rather more than originally asked for. When French visitors toured the production line at Voronezh in December 1972 eight Tu-144s were in final assembly, and parts had arrived for five more. Then, tragically, the second production aircraft suffered major wing failure at the Paris Salon on 3 June 1973, killing all on board. No report was published, but there had obviously been a violent manoeuvre, either to avoid a rogue Mirage or caused by a TV cameraman crashing on to pilot Mikhail Kozlov. Proving flights with cargo began on 26 December 1975, followed by passenger services on 1 November 1977. Subsequent experience was unsatisfactory. Regular services ceased on 1 June 1978, since when no Tu-144 has carried fare-paying passengers, though 16 production aircraft were flown, and got as far (in April 1982) as operating Aeroflot cargo services.

The final model, using almost the same airframe as the initial production version, was the Tu-144D. The major advance in this aircraft was to switch to the RD-36-51A engine, a totally new design by the bureau once named for V. Dobrynin, and now named Saturn. Rated at 44,090lb thrust, this engine had a large compressor with six rows of variable stators. Engine systems are grouped underneath, with access through doors under the nacelle, while airframe services such as pumps and generators are driven by a gearbox mounted on the airframe, inside the wing, driven by shafting and bevel gears from the HP spool. The afterburner was completely new, and featured propulsive nozzles with a translating conical spike to vary area and profile. It greatly reduced take-off noise, and was claimed to 'meet international

requirements for noise emission'. A total of 91 engines were made in 1968–78, and the crucial factor was that it enabled the Tu-144D to cruise at Mach 2 in dry thrust, reducing fuel burn from 26kg/km to only 11.2. Aircraft 77111 to 77116 were completed at Voronezh, but following a minor problem in May 1978 all Tu-144s were grounded. Then, unexpectedly, NASA opened negotiations with the Russian government for lease of a Tu-144D as a research vehicle to support the US HSCT (high-speed civil transport) programme. Aircraft 08-2, No. 77114, was overhauled and re-engined with four NK-321 engines of the type fitted to the Tu-160 bomber. Slightly modified, and civil certified at the same 55,077lb rating as the military engine, it powered 08-2 from 29 November 1996. This aircraft is stored for possible further use by a US agency.

Both the Concorde and the Tu-144 were underpinned by research aircraft. In addition to the Trident, *Gerfaut*, Griffon, various Mirages and Bristol 188, the Concorde was assisted by the HP.115 and BAC.221. The former was built specifically to explore the behaviour at low speeds (and thus high AOA) of the slender delta wing. Extraordinary in appearance, this aircraft comprised a wing of 20ft span and 430 sq ft area, which works out to an aspect ratio of 0.93! On the front was the cockpit, above the back the 1,900lb Viper turbojet, and underneath the fixed landing gear. Leading-edge sweep was just under 75°. As built, there was no droop or conical chamber, but the leading edge was detachable in order that different profiles could be explored. This aircraft first flew on 17 August 1961. The BAC.221 was a Fairey FD.2, which was completely rebuilt from 1961, flying in its new guise on 1 May 1964. The sharp-edge delta wing had slightly less span than its triangular predecessor, but area increased from 360 to 500sq ft. Investigation of vortex formation at low speeds was a crucial task, enabling full-scale results to be correlated with those obtained with models. The 221 had the advantage of being able to obtain results all the way from rest up to Mach 1.6 and back. It is typical of British officialdom that the FD.2 selected was the first of the two, the famous WG774, which had beaten the previous world speed record by 310mph in 1956. Now it no longer exists.

A Soviet counterpart was the MiG-21I, or A-144, mentioned earlier. Obviously, to be of value, a research aircraft must produce its information in time for the knowledge to be incorporated in the full-scale aircraft. With the BAC.221 the flow of information did not begin until the shape of the Concorde wing had been agreed, but at least this aircraft showed that the design was broadly correct. With the MiG-21I the timing was even worse. The job of conversion had been given to the MiG bureau, which was already overworked with fighter programmes enjoying the highest military priority. Work on the 21I proceeded at a snail's pace, with further delays caused by prolonged inability to settle details of the elevons, so that, by the time the *Analog* at last flew, the first Tu-144 was almost completed. Even if the 21I had been ready on time, four years in advance of the first Tu-144, it would not have helped in avoiding that SST's major problems.

In the United States, once the idea of a subsonic commercial jet had become respectable, in 1954, the concept of an SST followed not far behind. But there were two quite different routes to such an aircraft. One was the official, carefully studied view of the NACA (NASA from October 1958), which to some degree adopted an ivory tower outlook based entirely on the laws of thermodynamics and aerodynamics. This viewpoint accepted that the SST was going to be a project of awesome magnitude, and that it had to be a 'clean sheet of paper' aircraft without any compromises. The other route was to see whether any existing aircraft might be developed into an SST. There was one obvious candidate: the General Dynamics (Convair) B-58, although this was on the small side.

In 1959 I was privileged to have another long talk with Dick Sebold, who by then was Convair's VP

Engineering. He told me how straightforward it would be to develop the B-58A, which was then in full production, into the B-58C, and then develop this into an SST. The B-58C was to have essentially the same airframe as the B-58A, but instead of afterburning J79 engines it was to be powered by non-afterburning Pratt & Whitney J58s, similar to the basic core engines of the Lockheed Blackbird family. These were expected to give 65 per cent greater cruise thrust, with specific fuel consumption reduced by 35 per cent. To produce the B-58C-derived SST, a new fuselage was to be added, the wing centre-section spars being lengthened to match the greater fuselage width, the overall length going up from 96ft 9in to just over 155ft. A horizontal tail would be added, together with four sharply swept delta fins, or winglets, tilted two up and two down at 45° on the outer faces of the outer engine nacelles, which were centred on the tips of the wings. Sebold ventured the opinion that, provided the B-58C was fully funded already, they could then produce a competitive SST inside five years, and for much less than $100 million.

The problem was the crippling loss Convair was sustaining on the subsonic jet programme, the CV-880 and 990, which was to result in 1960 in the biggest single loss any company had ever suffered in one year. Had this not been the case, perhaps the Convair SST would have been built. But by 1960, partly because of the XB-70 and (for those in the know) the Lockheed A-12, Mach 3 had become strongly favoured. By this time even the nice, quick, cheap B-58C-derived SST had been

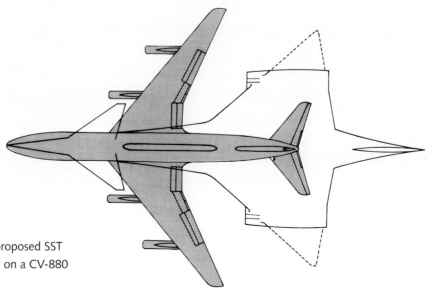

Convair artist's impression of the proposed SST based on the B-58C superimposed on a CV-880 subsonic jetliner.

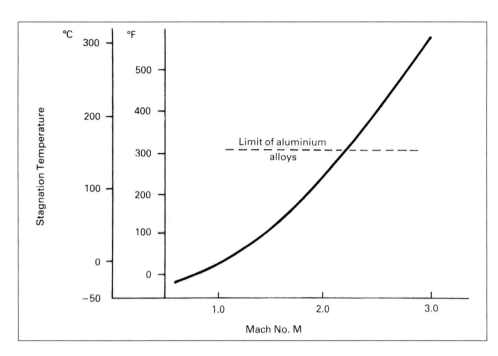

All supersonic flight results in significant heating of aircraft skin, and this becomes serious beyond Mach 2.

studied as a Mach 3 aircraft, with several significant changes including down-folding outer wings, as on the B-70, pivoted outboard of twin-engined pods, and the tailplane replaced by a canard foreplane. Later in 1960 Sebold gave evidence to the Congressional Committee on Supersonic Air Transports, in the course of which he showed a change in posture, in that he regarded the Mach 2 SST as 'an interim type'. The chairman asked: 'You propose to break this down into two steps. Instead of jumping from subsonic to triplesonic [which all other witnesses had taken for granted], you propose to go only to Mach 2?' Sebold answered: 'As a means of getting to the triplesonic.' I think by this time Convair had recognised that they were fighting a lost cause with a Mach 2 SST, despite its attractions of short timescale, moderate cost, and greatly reduced risk compared with Mach 3.

Whereas in Britain five years earlier those involved had adopted a hard-nosed realistic outlook based on cost, risk and operating economics, the Americans in the 1960 period appeared to adopt a 'nothing but the best' attitude. Anything less than Mach 3 soon came to be viewed as letting the side down. NASA played the central role, and, instead of concentrating wholly on costs and the eventual economics, its workers – all at Langley, except for one configuration studied at Ames – busied themselves almost wholly with how models behaved in the tunnel, and I mean a Mach 3 tunnel.

It all began when the British government freely handed over Barnes Wallis's Swallow concept, plus all the research data, in 1958. According to NASA, 'In 1960 the work on variable sweep was almost entirely confined to comments and tests on the Swallow concept.... for various reasons, the Swallow work was dropped in favour of a configuration with engines in the fuselage ... ' In 1958-9, five years later than in Britain, serious study began of SST prospects in the USA. This work culminated in late 1959 when a team from Langley summarised the situation in a Washington briefing for the then Administrator of the FAA, Lt-Gen Elwood 'Pete' Quesada. This was a time when everyone in the US industry was fired with enthusiasm for what variable sweep could do for the TFX programme (Chapter 10), and it was widely felt that this was the only way to achieve the desired L/D ratio, in both cruising flight and at take-off and landing. It even appeared possible to develop a cruise L/D 'much greater than that of the B-70', despite the fact that the bomber had been tailored almost entirely to the cruise regime.

In 1962 NASA Langley authorised the work on developing an optimised SST configuration to have full programme status. The US industry was happy to wait and see what answers came up, nobody committing any significant funds to in-house research for a further two years. Langley's Supersonic Commercial Air Transport (SCAT) studies embraced 22 configurations, but by 1963

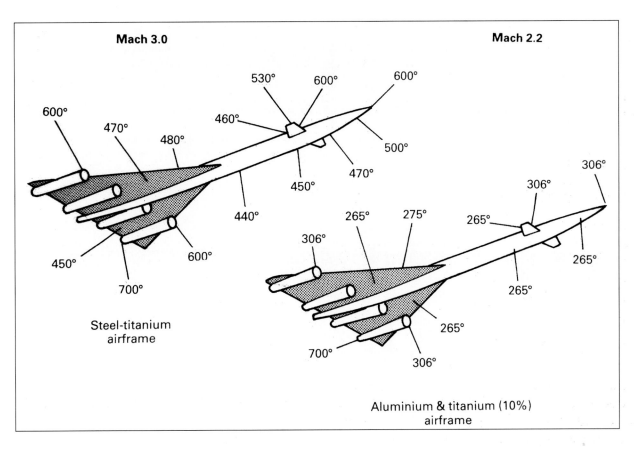

Simple diagrams showing skin temperatures for SSTs
cruising at M 2.2 and M 3. The degrees are in Fahrenheit.

these had been whittled down to four, and the following year to two. SCAT-16, a variable-sweep layout with wings of high aspect ratio pivoted to large gloves, and with a swept tailplane, was the basis for subsequent work by Boeing. SCAT-17, the only configuration originated at NASA Ames, featured a large fixed delta wing, initially plus a canard foreplane, and it formed the basis for subsequent work by Lockheed. NASA's contribution was enormous. Whereas in 1960 in the USA there was almost complete ignorance of SST design, by 1964 it was not only possible to predict aerodynamic efficiency and performance with precision, but all data could be fed to a computer program which, converted to punched tape, could be fed to machine tools which, in a few hours, could carve out a precise tunnel model.

Langley used hundreds of small aircraft models, some no bigger than a tiepin, in research into the physics of sonic booms. The same centre carried out extensive research into structures, especially into fatigue at Mach 3 temperatures. Other NASA centres carried out supporting programmes, notably Ames into aerodynamics and flight control, and Lewis into all aspects of propulsion. Lewis's work was partnered by industry. In 1957 Pratt & Whitney built two experimental engines, X-287 and X-291, to compete for the CPB contract which led to the XB-70. GE won this, as related earlier, but the X-287, or JT9A-20, was such a good engine it was later used as the basis for a range of possible SST engines, all of them advanced turbojets with afterburning thrust in the 35,000lb class. By 1959 the JT11B was on test. This single-shaft engine became the J58 for the A-12 and subsequent members of the Blackbird family. Its innovation was the use of six large pipes to bypass air from the compressor direct to the afterburner. Out of the JT9A and JT11B Pratt & Whitney developed an advanced turbofan with duct burning, which it proposed as its bid to power a future SST. Meanwhile, General Electric had in 1958 sent to the airlines a brochure describing a study engine called the X279M, a large afterburning turbojet with a single compressor spool with multiple variable stators. This engine was one of those picked by the USAF as suitable for a planned military SST, which seemed a sensible precursor to one for the airlines. Eight companies submitted bids for this Mach 3 aircraft, all with take-off weight exceeding 550,000lb and Republic Aviation's offering having ten engines!

In the event the USAF SST was never funded, though it provided valuable experience for the bidders. But in May 1963, while leaders of the industry were at the Paris airshow, President Kennedy unexpectedly announced that the United States – having gone to the Moon – would develop an SST. Management was vested in a special project office of the FAA, headed successively by Gordon Bain, Maj-Gen 'Bill' Maxwell (who had for five years headed the USAF SST), and finally former test pilot Bill Magruder. In 1967, when the US Department of Transportation was established, the SST programme was moved under its jurisdiction. The SST Program Office issued the specification to industry, assessed the submissions, and progressively whittled down the list of candidates. North American's NAC-60, with a fixed delta wing, small canard and bogie main gears retracting between each pair of engines, was eliminated in 1965. Lockheed's CL-823, with a double-delta wing of 'over 8,370sq ft' area, four individual engine nacelles and six-wheel bogie main gears retracting inwards, was eliminated on the last day of 1966. This left just the winner: Boeing's Model 733, with the SCAT-16 configuration. The winner for the engine, after a giant battle to convince the airlines that they actually made jet engines, was GE.

Their engine, the GE4/J4C, was very close to the original X279M of 1958. A large single-shaft variable-stator engine, it was rated with afterburner at 35,000lb. The four engines were installed in individual nacelles of circular section hung on pylons under the large fixed inboard wing. The inner engines were aligned with the longitudinal axis, and were hung on pylons canted sharply inwards, while the outer engines were toed inwards and hung on pylons canted sharply outwards, this odd arrangement enabling the pylons to pick up at the optimum points under the wing and still leave space between the nacelles for the forward-retracting bogie main landing gears. The outer wings, liberally endowed with slats, double-slotted flaps and spoilers, could pivot from 20° (173ft 4in span) to 74° (86ft 4in). Split flaps extended across the fixed wing. The vertical tail was sharply swept, and partnered by a ventral fin, while the horizontal tail was large and sharply tapered. The

beautiful fuselage had a constant diameter of 11ft 5in along the cabin, a length of 203ft 10in and a fixed nose. Gross wing area was 4,684sq ft, and estimated take-off weight (to carry 227 passengers across the Atlantic at Mach 2.7) 430,000lb.

The 733 looked to me a pretty good design. I doubt if many people, inside or outside Boeing, harboured any doubts that it would soon be translated into metal – mostly titanium. We shall never know if this would have been possible, because in early 1966, when the Lockheed rival still lived, Boeing threw away the 733 and started afresh. The company had been having agonising doubts over the aerodynamics, powerplant and, especially, the economics. Urgent discussions with the FAA and GE resulted in the latter scaling up the GE4 engine to an airflow of no less than 633lb/s, making the redesigned engine, the GE4/J5P, strictly the most powerful aircraft engine in all history. Features included advanced aircooled turbine blades designed to operate continuously at 1,150°C, hollow compressor blades to save weight, and the largest afterburner and con/di variable nozzle ever built. Take-off thrust eventually reached 69,900lb. (Some of today's turbofans have much greater take-off thrust, but their core airflow is nothing like 633lb/s and they are thermodynamically much less powerful). Development of the awesome J5P engine went ahead under John Pirtle in 1967, and the first of the uprated engines ran at 63,200lb on 25 March 1968. At this time the design take-off rating was 67,000lb, and the eventual figure of almost 70,000lb was a bonus. Engine dry weight was 11,300lb.

This great turbojet was matched to Boeing's completely new SST design, which was called the 2707. It abounded in novel features. The most remarkable was that the horizontal tail was greatly increased in size, and made a rearwards extension of the fixed inboard wing, so that when the outer wings were in their fully aft position (reduced from 74° to 72°) the fixed wing, movable wings and horizontal tail all merged to form a single huge delta wing with a straight leading edge and almost straight trailing edge, the tips being cut off at the Mach angle. The leading-edge flaps were continued along the fixed wing to the root. There was the expected twist and camber, giving the appearance from head-on of considerable dihedral. Even more startling was the mounting of the four engines, still in individual nacelles, under the horizontal tail. Thus, the tail had to be fixed, carrying just small elevons well inboard above

Boeing SST evolution from 1957 (Model 733-1) through 1971 (2707-300) showed dramatic rethinking on a scale unique in the history of aviation. Today everyone is agreed that the final configuration was right.

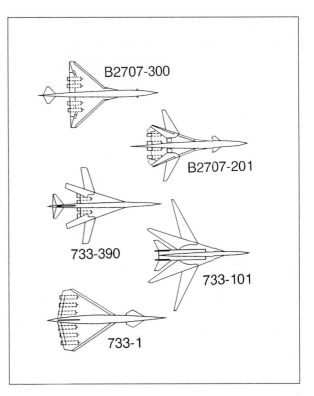

the engine nacelles. The vertical tail was redesigned, with taper rather than sweep and one section of rudder instead of four. At the front was a canard foreplane, with horn-balanced elevators, and long strakes downstream to generate vortices. Ahead of the canards the entire nose was arranged to droop, the front portion remaining horizontal, not because of high AOA but because of the poor forward view in the cruise position. Not least, the landing gear was redesigned, with two sets of four-wheel bogies in tandem on each side.

In some respects, such as the length of 318ft, the 2707 was the largest aeroplane of all time. Span increased to 105ft 9in (72°) and 174ft 2in (20°), wing area becoming 9,100sq ft and gross weight 675,000lb. As before, the fuselage was visibly area-ruled, there being no parallel tube portion and both height and width shrinking from the maximum at the leading edge of the wing. Seating arrangements were planned for 250 to 350 passengers carried as before at Mach 2.7 (1,800mph) at about 64,000ft for 4,000 miles Engineering and customer mock-ups were constructed, and the number of engineers assigned to the project swiftly grew to more than 2,000.

What nobody expected was that these 2,000 of the world's top designers would not be able to solve the problems. Ironically, while truckloads of technical papers appeared by engine designers eager to show that the optimum propulsion system was an augmented or duct-burning turbofan, or even a variable-cycle engine (as described later and in Chapter 7), the straightforward GE4 turbojet was the one thing about the 2707 that not only existed but also performed as advertised. The aircraft itself posed severe problems. The teams at Boeing, the FAA and, especially, at NASA, assigned to trying to solve the sonic-boom problem never really expected to come up with much, nor have they to this day. But the designers of the hardware did not expect to be defeated by, in the main, problems of aeroelasticity. Aircraft have to be flexible, but only up to a point. With the 2707 the interplay between the structural flexure, the flight-control system, the avionics software and the

sheer weight of metal in the aircraft, posed difficulties that, with the configuration adopted, proved impossible to overcome. There were also other shortcomings, such as the undesirability of putting the engine inlets exactly where they would ingest everything thrown up by the landing gears.

By late 1968 the aircraft had been redesignated 2707-200 to distinguish it from yet a third SST design which, to try to disguise the fact that it was a fresh start, was called the 2707-300. Boeing was naturally loath to give up variable sweep, which was the cornerstone of their design, and which had been by far the biggest factor in their win over NAA and Lockheed. With extreme reluctance, by August 1969 the decision had been taken to scrap the 2707-200, and set course afresh with the Dash-300, which was rather smaller, much simpler and, except perhaps in the matter of field length, generally superior. More to the point, it was capable of being built. Illustrations of all Boeing SST versions appear in another of my books for PSL, *Giants of the Sky* (1990).

The Dash-300 was a conventional tailed delta. The wing, of 141ft 8in span and 7,900sq ft area, had a t/c ratio of 3.7 per cent at the root, tapering to 3 at the tip, with 50° taper on the leading edge, increasing to 75° on the innermost portion. The four engine nacelles, almost unchanged, were now hung under the trailing edge of the wing. Between and beyond them were plain hinged flaps, the outermost portions operating as ailerons in low-speed flight. In cruising night, which remained Mach 2.7 at 60,000 to 73,000ft, roll control was by slot-deflector spoilers on both upper and lower surfaces. The entire 50° leading edge comprised plain hinged droop flaps. The vertical tail differed greatly from that of the Dash-200, while the sharply tapered, short-span horizontal tail resembled that of the original Model 733. The fuselage, reduced in length to 280ft, was area-ruled less obviously, though its upsweep towards the tail was made even more pronounced. The strakes on the forebody were retained, despite the disappearance of the canards, and the drooped nose was simplified to resemble that of Concorde. In front view the gull-winged appearance was striking, the wing having 4° dihedral to the outer engines and then 3° anhedral to the tips. All four engine nacelles were inclined downwards and toed inwards, and the main landing gears were changed to 12-tyre bogies (two pairs of twin-tyre wheels on three axles in tandem) mounted inboard of the inner engines and, seen from the side, almost in line with the engine inlets, where nothing could be thrown up into the engines. Structurally a major change was that the machined panels specified for the Dash-200 were all replaced by titanium sandwich skins, which tests showed were lighter and more rigid.

When Boeing redesigned their SST for the third time they had waited some two years for construction of a prototype to be authorised. Immediately after the last redesign, on 23 September 1969, President Nixon at last announced that two prototypes would be built, the first to fly in 1972 (soon changed to 1973) with production to begin in 1974. By 1970 production-line positions for 122 aircraft had been reserved by 26 airlines, but by this time a nationwide campaign was being whipped up dedicated to killing off the SST entirely. The reason was ostensibly the danger that would be caused by the SSTs to the environment. Many campaigners gained wide notoriety with their predictions of catastrophic sonic-boom damage, various lethal changes to the atmosphere, the dangerous effects of radiation on SST passengers, a wide range of extreme dangers caused by imagined faults in the aircraft, and pretty well anything else one cared to think of. Almost nobody had the courage to defend the SST. It was portrayed as a product invented solely to make money for the planemakers, which would benefit

nobody except 'a tiny handful of millionaires who callously care nothing for their fellow men'.

According to William A. Shurcliff, director of the CLASB (Citizens League Against the Sonic Boom), 'passengers and crew will be vulnerable to a number of potentially serious physical, physiological and psychological stresses associated with rapid acceleration, gravitational changes, reduced barometric pressure, increased ionizing radiation, temperature changes, and aircraft noise and vibration. Man cannot tolerate acceleration loads above 4 to 5 g. Visual disturbances occur between 3 and 4 g. At 5 g loss of consciousness occurs. Turbulent flight may cause brief linear acceleration of 10 to 12 g, which could cause fractures in unrestrained persons. Angular accelerations in turns and linear/angular accelerations during turbulent flight are important causes of motion sickness ... the exterior skin temperature will approach 260°C ... the strain on the SST Pilot will be enormous ... the SST is ... incompatible with other types of planes....' Shurcliff claimed that the collision between the F-104 and XB-70 was due to the fact that the bomber's pilot had 'poor visibility', ignoring the fact that the collision took place directly behind him, and even suggested an SST would never be properly tested, because 'the SST pilot, realizing that the prototype may be worth $250,000,000, must be careful not to make the extreme tests that would be of such interest'.

With hindsight, it seems beyond belief that such nonsense actually carried the day. Certainly a contributory factor was that Boeing, and to a much lesser extent GE, were deeply conscious of the problems, and knew there were easier ways to make a living. In late 1990 Don Bennett, a Director of Boeing international sales and once an engineer on the 2707, told me: 'Perhaps, if I had been more aware of the problems in economics, the environment and the politics, I would have worked on the SST with less enthusiasm. But today we can build a good SST.' I am grateful to him for reviewing this chapter.

According to GE: 'In 1970 a nationwide campaign was mounted warning of the dangers to the Earth's environment of SST operation. Following a series of debates arguing both sides of the issue in public forums, on TV and on the editorial pages of almost every newspaper and magazine in America, on March 29, 1971, the US Senate – by a vote of 49 to 48 – cancelled the SST program.' The 300ft mock-up became an attraction in a Florida amusement park, before passing through several other hands, being for a long period a major part of a church!

In a natural reaction to the loss of a gigantic programme, Boeing's Director of Advanced Transport Programs, Lloyd Goodmanson, told the 12th Anglo-American Aeronautical Conference in July 1971 that 'It is my opinion that a viable near-sonic transport could be in service before the end of the decade'. He gave delegates details of a Mach 0.98 aircraft, and of another cruising at 1.2, both with supercritical wings and area-ruled fuselages. In appearance they looked rather like an area-ruled 727 with two engines at the back and two under the graceful wing. Neither aircraft would produce a boom audible at the Earth's surface, but, though Goodmanson predicted competitive operating costs, he said reductions in drag and engine noise were needed.

During the 1970s research into suitable configurations for supersonic aircraft naturally continued, especially at NASA, and attention returned to one of the strangest configurations ever suggested for an aeroplane: the oblique wing. Apart from a German wartime project, the Blohm und Voss P.202 jet fighter, development of the concept is credited to Robert T. Jones, of NASA Ames Research Center. He reasoned that there was nothing

Dating from 1978, this SST was purely conceptual. It showed the sort of thing people were thinking of in order to use the slew-wing concept. The bodies were to be aligned side-by-side at low speeds.

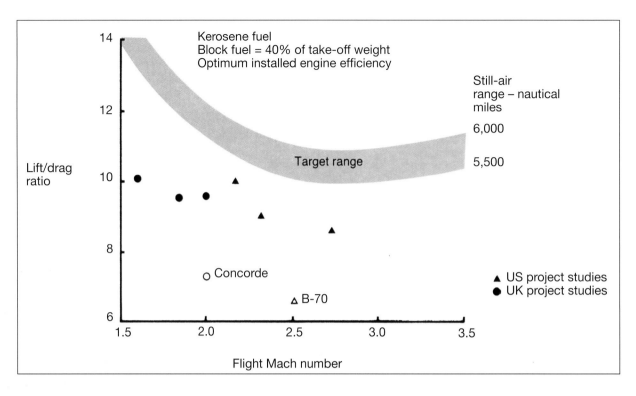

When one recalls the colossal effort that went into achieving optimum L/D for the Concorde and B-70, the progress made since seems scarcely believable. *Rolls-Royce*

intrinsically wrong with an aircraft lifted by a single wing pivoted in the centre, so that, as it was pivoted, one half became aft-swept and the other half forward-swept. This seemed so outrageous that some aerodynamicists even laughed, but tunnel testing showed that such aircraft could indeed fly. In 1978 NASA awarded a contract for the AD-1 research aircraft to Ames Industrial Corporation, the detail design being carried out by Bert Rutan. The AD-1 first flew on 21 December 1979, and did all it was supposed to do. Powered by two tiny turbojets of 220lb thrust each, it was not designed for speeds much in excess of 200mph, but its 12 per cent wing could pivot from zero (i.e. at 90° to the fuselage) to 60°. The oblique wing was not specifically associated with supersonic flight, but, according to NASA Ames and also NASA Dryden (where the AD-1 was tested), the concept 'offers major advantage in the design of an SST'. It was claimed to offer 'a fuel economy twice as good

as that of the first SST generation … and a substantially weaker sonic boom'. NASA has done a fair amount of tunnel testing on oblique-wing transports, and has even briefly looked at the idea of a twin-fuselage version, the two bodies being linked by both wings and tailplanes, but we haven't heard much in the past 20 years.

By 1990 work was gathering momentum on an *Avion de Transport Supersonique Futur* (ATSF), or Concorde II, or Advanced Supersonic Transport (AST), or High-Speed Commercial Transport (HSCT). When I recall how in the 1960s we seemed to be banging our heads against a wall – and I don't mean the wall of sound – it is truly remarkable how the past 45 years have transformed the picture. In structure weight percentage, L/D, ratio and, above all, in the propulsion system, the advances in technology guarantee that an eventual 'SST2' would be a different species, with dramatically better economics and meeting all environmental legislation. The one thing nobody is banking on is supersonic flight over land (Aérospatiale says 'over residential areas', but I think they are optimistic). In 1992, when I wrote the first edition of this book, I expected a second-generation SST2 to be in service by 2010. What I did not expect was that, by 2008, the enthusiastic SST2 teams would almost all have been dispersed, so that – almost for the first time – the current *Jane*'s omits any mention of such

a programme. However, for completeness I will repeat the prospects as they looked in 1992.

For a start, there is almost total unanimity on size of aircraft (230–280 passengers), cruise Mach number (2.2 to 2.4) and configuration (four engines individually spaced under the trailing edge of a wing of what is often called double-delta shape). This means that there can be a lot of co-operation, and an unprecedented amount of effort applied to solving the problems and refining a single final design.

The most fundamental advance is that of the basic vehicle aerodynamics. In the years immediately following

Artwork showing an early configuration of the Avions de Transport Supersonique Futur.

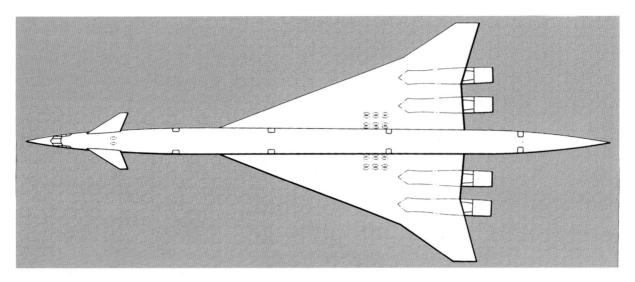

British Aerospace AST, mid-1990.

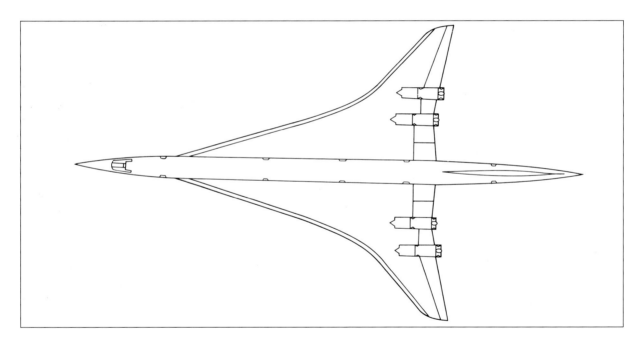

Aérospatiale ATSF, mid-1990.

the Second World War aerodynamicists were looking for supersonic L/D ratios around 4. By the early 1950s wings had improved enormously; the Supermarine 545 wing looked superficially very like that proposed for the future SST, and achieved L/D of 6 at Mach 1.6. Concorde achieved 7.2 at Mach 2, but since then prolonged research, assisted by computers, has led to advances many times greater than anything dreamed of when

Concorde was designed. All the companies working on a future SST have achieved L/D greater than 9.8 at Mach 2 to 2.25, and there is no doubt that when such an aircraft is built it will cruise at an L/D slightly in excess of 10. This is a marvellous achievement for a supersonic aircraft.

The gains have come partly from improved profiles, partly from a better planform, and partly by painstaking reduction of drag. A remarkable unanimity exists between all of today's SST teams. All are agreed on a fixed-geometry wing of so-called double-delta form, with a large inner portion with a leading-edge sweep of 70–75° and a more conventional outer section swept at 40–50°. Having said that, there are still differences. British Aerospace has the smallest span (115.5ft), the smallest and most sharply swept outer sections, with sweptback trailing edges, and most pronounced anhedral over the outer panels. Aérospatiale has a beautiful wing, with gentle curvature joining the inner and outer sections, the biggest contrast in sweep (75° inboard and 42° outboard), the greatest span (140ft) and a leading edge curved in towards the bottom of the fuselage. Boeing has even greater leading-edge sweep towards the root, an intermediate span (119ft 10in) and, unlike the Europeans, a trailing edge swept forward from root to tip.

All designs have trailing-edge elevons, the spacing from root to tip across the two engines of each wing being 2-2-3 for BAe, 2-1-3 for Aérospatiale and 1-1-2 for Boeing, the latter also having outboard spoilers. A major advance in comparison with Concorde is that all envisage leading-edge flaps, or conceivably slats. These in no way negate the achievement of additional lift from leading-edge vortices. One BAe artwork shows a non-canard AST in the approach configuration with giant hinged surfaces resembling inverted Kruger flaps open, above the inboard leading edges. But, apart from this, there are more basic differences between the three published designs. Aérospatiale proposes a tailless delta configuration, relying as in Concorde on elevons not far aft of the CG. BAe has a similar layout, plus a canard foreplane, the entire surface powered, mounted high on the fuselage immediately aft of the flightdeck. This gives 'significantly improved control and lift'. Boeing proposes a large tailplane, as in its 1971 design.

In the fuselage there are surprising differences. BAe suggests a constant width but variable height, to give an area-ruled profile. The cross-section is almost circular, one computer image showing an interior width of 156in (13ft) and aisle height of 99.93in. Aérospatiale favours a constant cross-section, almost identical to BAe's but without variation in height, achieving area ruling by the longitudinal distribution of the wing. Boeing proposes considerable area-rule variation in both width and height, together with the upward curve at the tail seen in 1971. Despite the canard, BAe would land nose-high, and expects to use a droop-snoot hinged nose. Jacques

Plenier, Manager of Aérospatiale's Aircraft Division, told me, 'I am sure we will retain a nose similar to Concorde, though we would like to make it fixed.' Boeing's view is the same: a droop snoot is a necessary evil, even with a tailplane. All three companies propose main gears which in side view have three axes in tandem, but similarity ends there. BAe's gears are hinged ahead of a point midway between the engines, so the entire gear has to be stowed in the wing, which means 12 small tyres on each leg. Aérospatiale hinges its legs inboard of the inner engines, as on Concorde, so it can use six big tyres on each leg, all housed in the fuselage. Boeing proposes four separate main gears, each hinged ahead of a point just inboard of an engine; all four must be housed in the wing, so each has three twin-tyre wheels in tandem. Of course, everyone agrees on digital bus-connected avionics, with fly-by-wire (FBW) flight controls.

Compared with Concorde, which is for everyone the baseline aircraft against which to measure improvements, the gains made in each area look impressive. For example, the improved cruise L/D promises an increase in range of some 35 per cent, extended to 45 per cent by improvements in engine specific fuel consumption (10 per cent is a conservative gain here). Further modest improvements in L/D, all known and almost certainly attainable, promise to extend the range increase to 60 per cent. Advanced structure and materials, notably high-temperature composites, aluminium-lithium alloys and metal-matrix structures, promise to save at least 30 per cent by weight, increased to 40 per cent by improved design; this is equivalent to saving 6.5 per cent on take-off weight. Studies by Rolls-Royce show that, for a given thrust, installed powerplant weight can be reduced by one-third, giving a further 4 per cent reduction in MTO weight. VC engines will be more efficient in subsonic flight, cutting reserve fuel for subsonic flight, diversion, and holding from 7 to 5.6 per cent. Adding these gains together gives a reduction of 12 per cent in MTO weight, which is equivalent to multiplying payload by about 3.2. Tripling the payload, and increasing range by 60 per cent, means 4.8 times as many passenger-miles per trip, which puts a future SST in contention with today's subsonic airliners.

In passing, as one can get even better aerodynamic and propulsive efficiencies at Mach 3 to 3.5, which is what the US aimed at in the 1960s, why lower the sights to 2–2.5? The answer is one of common sense and mathematics, quite apart from any 'gut feelings'. No aircraft is worth creating if it cannot be sold, and tomorrow's SST has to be not only safe, and environmentally certifiable, but also economically attractive. At Mach 2.2 it is possible to create an aircraft that meets all the requirements, could operate at seat-mile yields around 10 cents (compared with 87 for Concorde) and, priced at about 500 million 1990 US dollars [I repeat, this was written in 1992], could command a market of some 400 aircraft. What limits the number sold is the aircraft's own productivity. Sir Ralph Robins [then chairman of Rolls-Royce] said: 'The problem with a next-generation SST is that its productivity is so tremendous that nobody wants more than one or two.' Going to Mach 3 or more multiplies the productivity by 1.5, thus dividing the number needed by the same factor, while simultaneously at least doubling the already colossal development bill. For a Mach 2.2 SST the bill is tentatively put at US$10 billion. The whole project makes economic sense, whereas Mach 3plus does not. Thus, Aérospatiale's AGV, described briefly in Chapter 12, is in my opinion pure fiction.

We have already seen that there is no doubt that the future SST can have a range at least 60 per cent greater than that of Concorde. The published figures in late 1990 were interesting. Sid Swadling, Director of Engineering of BAe Airbus Division, says, 'AST will fly 5,500 nautical miles (6,330 miles) compared with 3,500'. Jacques Plenier says, 'ATSF doubles the range, from 3,200 n.m. to 6,500 (7,475 miles)'. Lawrence W. Clarkson, Boeing Senior VP for International Affairs, says, 'Originally we were looking at a 3,500 n.m. minimum, but every day the demand increases, and I believe we will eventually have to deliver 6,000'. Another thing that happens every day is the growing importance of the Pacific Rim in terms of air traffic. The Pacific is ideal territory for an SST, and a 6,000 n.m. (6,900 mile) sector distance suits this market beautifully. It also matches the UK–Australia route with one stop (Colombo), taking 12.2 hours compared with the 20.7 hours of a heavily laden non-stop 747-400.

Such an aircraft would have a takeoff weight of about 650,000lb, 50 per cent up on Concorde. Aérospatiale published an ATSF brochure quoting 485,000, but in 1990 Plenier agreed that this was optimistic. Malcolm MacKinnon, Boeing's HSCT design manager, concurs with BAe. Such an aircraft would require four engines, of the kind discussed in Chapter 7, each in the 60,000lb class. Overall length figures are: AST, 300ft; ATSF,

The Boeing HSCT compared with a 747-400.

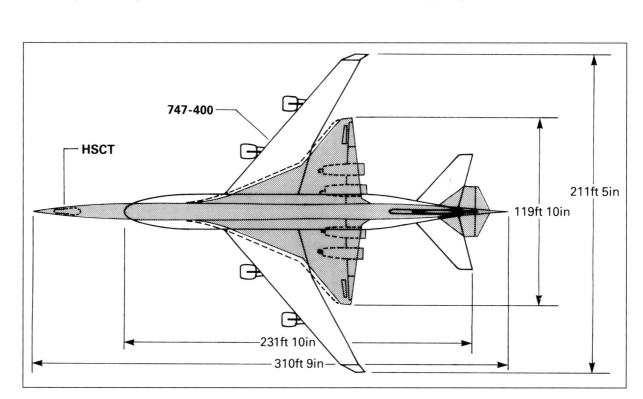

A rather stylised take-off by the BAe AST.

249ft (bound to increase); HSCT, 310ft 9in. Seating capacities in 1990 were: AST, 275–300 (289 at uniform 34in pitch); ATSF, 200 (two-class) to 250; HSCT, 280 (three-class).

In the early 1980s BAe and Aérospatiale began to spend significant money studying SST2. As their thinking was so similar they decided to work jointly, pooling information and reaching agreement on transfer of technology gained since the end of the Concorde programme. But almost everyone agrees that there is room for only one SST2, and in May 1990 agreement was reached linking five companies in an initial one-year study. The participants are Aérospatiale, Boeing, BAe,

Deutsche Airbus and McDonnell Douglas. This group is not concerned with detailed engineering, nor with pooling design data, but with broad issues. Two working groups were set up, one to study business matters, and the other technical and marketing subjects. Two subjects at the top of the agenda are environmental issues, and the likely requirements for SST2 certification.

When the year is up, in 1991, I do not doubt that the partners will move on to the next stage. I hope they will be joined by the Soviet Union and Japan. BAe's Swadling said, 'Concorde moved into new areas of technology. AST will not do that, so that the gestation period will be correspondingly shorter.' Despite this, nobody is exactly rushing to get the aircraft into the sky. Most people are thinking in terms of 2000–2005, for airline service in 2010–2015.

General arrangement of the Gulfstream/Sukhoi SSBJ.

That is what I wrote in 1991, and the 'view from the bridge' in 2008 is very different. I never expected the fastest passenger vehicle to be withdrawn prematurely, with no thought of a successor.

Over 50 years ago a project to modify a Vampire fighter into a 4/6-seat passenger aircraft launched the idea of the business jet. Today such aircraft serve around the world in thousands, proving the old adage that time can be equated with money. Since the early 1960s some designers have wondered how far the process can be taken. Ed Swearingen was one of the first who studied the prospects for a supersonic business jet (SSBJ), which at the cost of a lot of money could save a lot of time.

Surprisingly, the first team to take positive action in this field was the OKB named for P. O. Sukhoi in the Soviet Union. Under General-Constructor Mikhail P. Simonov, a team led by Aleksandr Blinov began work in 1987 on a project given the bureau number S-51. As is commonly the case with major Soviet aircraft, the basic shape was decided in collaboration with the

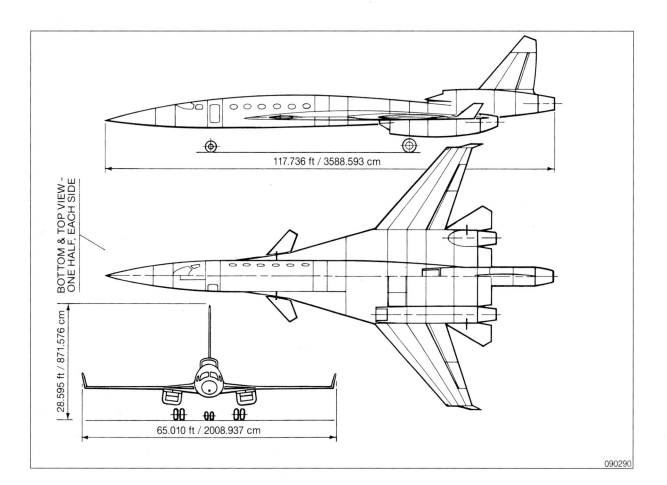

117.736 ft / 3588.593 cm

BOTTOM & TOP VIEW - ONE HALF, EACH SIDE

28.595 ft / 871.576 cm

65.010 ft / 2008.937 cm

090290

Central Aero-Hydrodynamics Institute, led by Guerman (Herman) I. Zagainov. The result was an area-ruled bizjet with a low-mounted swept wing of quite high aspect ratio, engines (probably two) at the tail, fed from a variable vertical inlet of F-107 pattern above the fuselage, and, strangely, no horizontal tail. Seating was to be provided for 12 to 21 passengers, '… from any capital city to any other at Mach 2 with never more than one intermediate stop'.

At the 1988 Farnborough airshow Gulfstream Aerospace announced that it was carrying out an engineering study on an SSBJ, while Rolls-Royce disclosed that it was studying the propulsion requirements for this aircraft. In June 1989 Mr Simonov reaffirmed what he had told the author at Paris, that his OKB was seeking foreign partners. He particularly mentioned Gulfstream, saying: 'With their work on business jets, and our experience of supersonic aerodynamics, we could work well together.' He then met Gulfstream's Allen Paulson, and executed

Gulfstream/Sukhoi SSBJ. Studies continue, though no longer with Sukhoi as partner.

a Memorandum of Understanding. Paulson told the Gulfstream Operators' Workshop, 'The Sukhoi bureau appears to be years ahead of the rest of the world in the design and development of an SSBJ.' The whole project really got off the ground at a big meeting in Moscow, at which Rolls-Royce and Lyul'ka got together on the propulsion of what soon took shape as an attractive tri-jet. This later metamorphosed into a twin-jet, and then in 1992 Gulfstream withdrew. Sukhoi struggled on for a while but eventually gave up. Gulfstream kept thinking, and in 1998 revealed that it had reached agreement with Lockheed Martin to work jointly on a M1.6-M2 aircraft, unrelated to that studied with Sukhoi, 'similar in size to a Gulfstream II'. By the new century this had become

151

the Quiet Supersonic Jet (QSJ), weighing 100,000lb and cruising at M1.8, and with a long nose spike which from August 2006 has been tested on the NASA F-15B No. 836.

Meanwhile, a new kid has appeared on the block. In 2002 Aerion Corporation, based in Reno, Nevada, announced that it had been formed 'to reintroduce commercial supersonic flight'. It has published artwork showing a straight-wing aircraft of conventional configuration, to be powered by two 'new variants' of the Pratt & Whitney JT8D-219 turbofan, which in its original form was a subsonic engine rated at 21,000lb. Though very 'long in the tooth', this engine is to go back into production in 2010 to re-engine the USAF Boeing E-8 Joint-STARS fleet, which would certainly assist this civil application. Seating up to 11, in a 93ft fuselage, the Aerion is to have natural laminar flow over its wings and horizontal tail. Aerion say they have Letters of Intent for more than 40 SSBJs, at a planned price of $80 million. It remains to be seen whether they can make good on their intention to certify this aircraft in 2014.

Aerion's Supersonic Business Jet for eight passengers will have a cruising speed of Mach 1.5 and a range of more than 4,000 nautical miles. The US-based Aerion Corporation claims it will be possible to cross the United States at Mach 0.98 with operating costs equivalent to today's large business jets. Certification is expected in 2014. *Aerion Corporation*

Chapter 10

V/STOL AND VG

In the late 1950s two things happened that introduced a gigantic hiccup into the development of combat aircraft. One was jet lift, leading to vertical or short take-off and landing (V/STOL) aircraft. The other was variable geometry (VG), which in this context means pivoted variable-sweep wings, colloquially called 'swing wings'. Neither idea was new; they just happened to reach the stage of practical application around 1960. An objective observer would quickly have come to the conclusion that they would do more than cause a mere hiccup. Both ideas can be shown to result in superior warplanes, VG because the pilot can redesign the aircraft in flight to match it to totally different flight regimes, and V/STOL because only this species of aircraft can operate without going near known airfields, which in war against any but the most primitive enemy would be wiped off the map by nuclear attack in the first few seconds.

Strangely, there is a lot of fashion in aviation, especially if the new technology happens to be invented somewhere else. For at least 20 years jet lift and variable sweep have been out of fashion. Supposed experts talk only of 'penalties', either in flight performance or in structure weight. Clearly, they prefer aircraft which might be fractionally lighter or simpler or faster, even if they are wiped off the map before they can get into the air to show how good they are. Indeed, in the twenty-first century one might be forgiven for thinking that nuclear weapons had never been invented, because the idea of dispersing assets away from known airfields has been forgotten.

The technology of powered-lift V/STOL was explored by numerous research aircraft in the 1950s, many of them weird and all highly subsonic. The first practical scheme was devised by Hooker at Bristol (engine) and Camm at Hawker (airframe), using an engine with four vectoring nozzles installed so that these nozzles straddled the aircraft CG. This simple arrangement survives today in the Harrier family, which would be the only warplanes likely to keep operating on day two of any genuine war. But these are subsonic aircraft, and in 1960 the notion of a subsonic tactical aircraft was regarded – by the RAF, at least – as some kind of insult. The then Chief of Air Staff said to me in 1960, 'I hope you aren't one of those people who actually expect us to *buy* the P.1127?' The P.1127 was the original Hawker vectored-thrust prototype, and to be fair it gave little hint of what it might be developed into. Realising the situation, Hooker quickly came up with a developed engine with plenum-chamber burning (PCB), a form of afterburning, in the front nozzles. Camm then used this to plan the P. 1150, a V/STOL fighter/attack aircraft to reach Mach 2.1. Much longer than the P.1127, and with its vectored nozzles streamlined in the lee of the variable supersonic inlets, it looked like the start of something big.

Meanwhile, the NATO staffs at SHAPE recognised that in powered lift they had the answer to the problem that had worried them desperately from the mid-1950s: how to disperse Allied air power so that it should not be destroyed by Soviet missiles before they even knew they were at war. Final agreement on what seemed to be a crucial specification for a V/STOL tactical aircraft was reached in March 1961. NATO Basic Military Requirement No. 3 (NBMR-3) initially asked for the ability to operate from a 200m (656ft) unpaved strip at maximum weight, carrying a 2,205lb bombload over a lo-lo-lo radius of 288 miles. Speed was to be not less than Mach 0.92 at sea level, and not less than 1.5 at high altitude. In August 1961 the requirement was refined to add the ability to clear a 49ft screen 492ft from brakes-release.

The result was a deluge of interesting projects from almost every maker of warplanes in the NATO nations. Some used vectored thrust, while others preferred Rolls-

The Dassault Balzac V 001 in tethered hovering flight.

Much larger than the Mirage III, the IIIV was lifted by eight RB.162s with upper doors quite unlike those of the Balzac. This was IIIV.02, which achieved Mach 2.04.

Royce's so-called 'composite' formula of a battery of special lift turbojets used only at take-off and landing. It was easy to show that each idea was better than the other. One proposal, the Fokker-Republic D.24 *Allianz*, combined vectored thrust, lift jets and swing wings. At Kingston, Camm discovered that there was no way he could stretch the P.1150 to meet the requirement. Under intense pressure, he began work on a fresh design, the P.1154, while Hooker rushed to produce the BS.100 engine for it, in the 33,000lb class. It was soon obvious that the vast NBMR-3 competition had become a two-horse race, and in April 1962 the assessment team announced that the P.1154 was 'technical first choice'. But, in order not to offend the French, they were forced to declare that the Dassault Mirage IIIV was 'of equal merit'. In the longer term, NBMR-3 just faded away, and NATO air power has ever since been parked on what were, until the collapse of the Soviet Union, the most heavily overtargeted spots on this planet. But two national V/STOL programmes remained.

The P.1154 had succeeded in breaking through the 'no more manned aircraft' rule, and on 18 July 1962 it became a firm national programme, initially to replace the Hunter in the RAF and the Sea Vixen and Scimitar in the Royal Navy. It was unfortunate that this happened just as US Defense Secretary McNamara was insisting on a common TFX (see later in this chapter) for the USAF and Navy, because naturally the idea mesmerised the British politicians. Equally predictably, the RAF's air marshals and the Royal Navy's admirals went out of their way to make their versions of the P.1154 as different as possible. Eventually, in 1964, the admirals dropped the P.1154, and picked a carrier-based version of the US-built Phantom, and a year later the newly elected Labour government cancelled the British carrier force, ensuring that the Phantoms could not go to sea. This left only the RAF version, and in February 1965 the new government cancelled that also, again buying a special version of the Phantom, ensuring that the RAF would remain tied to those overtargeted airfields.

As a small concession, the RAF was allowed to buy a developed version of the P.1127, and this eventually became famous as the Harrier. Without it the Falklands would still be part of Argentina, but being subsonic the Harrier does not feature here. In France, however, Dassault were building the Mirage IIIV to fly at Mach 2. It was being strenuously promoted not only by Dassault but also by Boeing, the British Aircraft Corporation and Rolls-Royce. The latter firm, with an eye to sales of thousands of lift engines, maintained that the single-engined concept, exemplified by the P.1154, was inefficient. In contrast, the composite solution enabled the cruise engine to be smaller, sized exactly to the propulsion requirements of the aircraft, and thus burning less fuel. They glossed over the fact that the aircraft would be burdened at all times by the enormous bulk and not insignificant weight of a battery of lift engines, used only for take-off and landing. Dassault began with the Balzac, a converted Mirage III with eight Rolls-Royce RB.108 lift engines and a Bristol Siddeley Orpheus for propulsion. It was one of the few aircraft to crash fatally, be rebuilt, and then crash fatally again. It led to the much larger Mirage IIIV, lifted by eight RB.162s and propelled by a Snecma TF 106. Two Mirage IIIVs were built, the first being hovered on 17 February 1965. On 25 March 1966 it made a successful transition from hover to wingborne flight, and back to the hover. The second IIIV was flown in June 1966, and on its eleventh flight, on 12 September of that year, it reached Mach 2.4, roughly twice as fast as any other VTOL aircraft before or since. But the whole programme was abandoned, largely because the vulnerability of airfields – and thus the uselessness of conventional airpower – was swept under the carpet and ignored.

Until the F-35B began testing in 2008 the only supersonic V/STOL aircraft have been complex designs which failed to achieve their objectives. The German EWR-Süd VJ 101D was to be a Mach 2 interceptor bearing a faint resemblance to the F-104, with two RB.153-61 augmented turbofans for propulsion and a row of five RB.162s for lift. It was never built, but much flying was done with VJ 101C research aircraft with paired RB.145 turbojets in tilting nacelles on the wingtips, plus two lift engines in the fuselage. This seemed an attractive solution, because the aerodynamic and structural penalties were small. One VJ 101C reached Mach 1.04, but, with jet-lift going out of fashion, the programme was abandoned. Details appear in *Rolls-Royce Aero Engines* (PSL, 1989).

Rolls-Royce and the famed Indianapolis-based firm of Allison collaborated on a brilliant third-generation lift jet, the XJ99, which typically gave 9,000lb thrust for a weight of 440lb. Two pairs were to be mounted in complex swing-out doors on each side of the fuselage of the EWR-Süd/Fairchild Republic AVS (Advanced Vertical Strike), which was also known as the 'US/FRG'. This had everything, including deflected main engines

Artist's impression of the NAA XFV-12A operating from a ship's deck. In the vertical mode the entire engine jet was piped to lifting ejector nozzles along the wing and foreplane. *Philip Jarrett*

The XFV-12A is something Rockwell might prefer to forget. *Philip Jarrett*

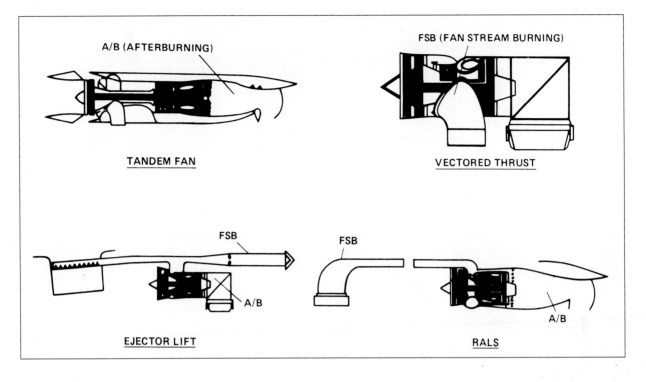

A/B (AFTERBURNING)

TANDEM FAN

FSB (FAN STREAM BURNING)

VECTORED THRUST

FSB

A/B

EJECTOR LIFT

FSB

A/B

RALS

and pivoted swing wings, but it was cancelled before first flight in 1968. Today Allison is Rolls-Royce North America, playing a central role in the F-35B.

An even more problem-ridden aircraft was an American attempt to demonstrate a V/STOL far more advanced than the derided Harrier. The job was given to North American Aviation, whose NR-356 design was accepted by the US Navy in late 1972 as the best submission for what became the XFV-12 fighter/attack aircraft. This was to operate from small flight decks, and possibly even from helicopter platforms on surface warships, rising vertically with full weapons load and then accelerating to over Mach 2. Features included a rear-mounted wing, a large canard just forward of the CG, twin vertical tails, and a single Pratt & Whitney F401 augmented turbofan. The clever part was that, in the vertical mode, all the engine efflux would be diverted by a valve upstream of the afterburner, and piped to numerous downward-pointing nozzles in pipes running from root to tip inside the wing and canard. Except for the leading edge, the entire upper and lower skins of these surfaces were in the form of hinged panels. When open, air induced downwards through the resulting ducts by the ejector nozzles provided the lift. It looked a formidable combination of clever ideas, but unfortunately it failed to perform as advertised.

Simple diagrams showing the four main methods of making a single engine give lift and propulsion. The multinational ASTOVL programme cut short the only one known to work, vectored thrust.

Since then the whole idea of jet-lift V/STOL (or rather STOVL) has been put on the back burner. Today nobody believes that any enemy would be so unsporting as to eliminate airfields with missiles, so the urgently needed survivable warplane became regarded as a vague prospect for some time in the future. Companies in the UK, USA and Canada have since January 1986 collaborated with NASA, the RAE and user services on a programme called ASTOVL (Advanced STOVL). It was agreed that a future aircraft would have a single lift/cruise engine, outstanding agility and a peak Mach in excess of 1.6. Four candidate systems were identified: advanced vectored thrust (AVT), augmentor ejector, remote augmented lift system (RALS) and tandem fan.

I was fortunate enough to be a speaker at the December 1987 Powered Lift Conference in California, and it was clear that, except for the British, nobody liked AVT. The fact that it was the only system known to work carried little weight and, whereas in the complex rival systems delegates extolled the virtues, with simple vectored thrust they merely harped on the drawbacks or supposed difficulties. A year later this was the first scheme to be rejected, which I thought a mistake. This left RALS, in which fan air is ducted to an auxiliary burner and nozzle near the nose; augmentor ejector, with the bleed air (or possibly the whole mass flow) ducted to banks of ejector nozzles inducing a much larger downwards lifting airflow; and tandem fan, or hybrid fan, in which for vertical flight doors are opened to allow the front fan to blast downwards, roughly doubling the engine

airflow, instead of feeding the rest of the engine. The ejector scheme has come a long way since the XFV-12, and has been especially developed by NASA, General Dynamics (using modified F-16s) and Boeing Canada, but attainment of the theoretical results was in 1990 still proving elusive. The tandem fan has much in common with the VC engines needed for future SSTs.

When I wrote the previous edition, only the Soviet Union had a new jet-lift STOVL. This was designed by the Yakovlev bureau in 1975, a ship-based air-defence aircraft, with the service designation of Yak-41. More than 50 configurations were studied, before deciding to use a slightly swept folding wing mounted on top of a large box-like fuselage containing a single afterburning turbofan, a Soyuz R-79V-300 rated at 34,170lb. The propulsive nozzle was made of tapering rotating sections so that it could pivot through 95°, to give lift, forward thrust or braking. Even with the tail carried on left and

The Yakovlev 41M was the first supersonic jet lift aircraft. Here it is hovering, with the lift-jet doors open and main-engine nozzle at 95°. *PRM Aviation*

right girders, the nozzle was still well behind the CG, so two RD-41 lift jets, each giving 9,040lb upthrust, were mounted between the main engine and the cockpit. Funding covered four Yak-41Ms, two of them fully equipped flight-test aircraft. The first, 75, took off down the Zhukovskii runway on 9 March 1987, and the other, 77, made the first hover on 29 December 1989. Peak Mach number reached was 1.74, and many records were gained, submitted to the FAI under the invented designation of 'Yak-141'. This number was later painted on both aircraft, misleading many Western writers.

The collapse of the USSR effectively terminated the Yak-41M programme, while at the same time the USA was ready to launch a project for an all-new STOVL fighter. In November 1994 the US Naval Air Systems Command merged several predecessor programmes into the JSF (Joint Strike Fighter). Boeing proposed the X-32B, which looked like no other aircraft before or since. It featured a flattened fuselage, which merged into a broad wing carrying sloping fins on the trailing edge. A large chin inlet fed air to a large augmented turbofan engine, which in normal flight exhausted through a conventional afterburner. For vertical lift the flow was switched downwards through Harrier-type nozzles, at the CG, while large secondary flows blasted through yaw, pitch and roll nozzles. Lockheed Martin proposed the X-35B, which was originally based on the Yak-41M. In the vertical-lift mode this swivelled its afterburner down in the same way, but instead of having forward lift engines the main engine could be coupled to a shaft putting 29,000 shaft horsepower into a pair of large contra-rotating fans in a vertical duct behind the cockpit of a fairly conventional-looking fighter.

Following competitive tests with prototypes powered by F119 engines (next chapter), the choice fell on the Lockheed F-35 Lightning II, powered by the Pratt & Whitney F135, rated at 40,000lb thrust. A so-called 'alternate' engine is the slightly later and more powerful General Electric F136, in partnership with Rolls-Royce.

The first F-35A prototype flew on 24 October 2000, and the first production F-35A on 15 December 2006. In April 2008 a nominated British pilot tested the engine of the first F-35B, the STOVL version, ready for the start of flight-testing later in the year. The F-35C is the carrier-based version, with many airframe differences. For almost a decade impressive lists have been published of proposed F-35 sales to many countries, including 1,763 for the USAF, 680 for the Navy/Marine Corps, 90 for the RAF and 60 for the Royal Navy.

Today we think of jet lift in a purely military or naval context, but nearly 60 years ago A. A. Griffith, a visionary who worked at Rolls-Royce, dreamed up

One of the Griffith Swallow variable-sweep models on test in 1958 at NASA Langley, after the technology had been passed to the USA. Of course, the engine pods had to swivel as well.

Opposite top: The Grumman F-14 was almost everything the F-111B was not, apart from having to use the same engines. The mock-up, shown here, was originally built to Configuration 303E, with a single fin and twin folding ventral fins.

Opposite bottom: Lockheed F-35 Lightning II. This is the STOVL version, seen with lift-jet doors open. *Lockheed-Martin*

VTOL SSTs! His employers liked the idea, because each aircraft was going to have upwards of 40 engines. One illustrated in *Rolls-Royce Aero Engines* had 45, and a later version had 58. Batteries of light and compact lift jets, such as the RB.162, would lift off a shape optimised for Mach 2.2, and quite unable to take off normally. Propulsion jets were to be housed in the fins. The idea took a long time to go away. In the 1970s Hawker Siddeley had many projects for V/STOL transports, in military, airline and executive categories. Most were subsonic, but for interest I have included one of the jet-lift SSTs from 1976. It would have been lifted by batteries of relatively quiet high-bypass turbofans arranged along large pods inboard of the underslung propulsion engines. I think that today such things are non-starters.

In Chapter 4 I mentioned that in 1942 Alex Lippisch filed a patent for the concept of a wing mounted on a pivot so that its angle of sweep could be varied. In 1943 Messerschmitt AG began the design of a completely new jet fighter, the P.1101, the first project to have swept wings. These were made adjustable on the ground between 35° and 45° so that the best angle could be determined by trial and error. As related in Chapter 8, this led to the Bell X-5. The next VG aircraft was the Grumman XF10F Jaguar carrier-based fighter, first flown in May 1952. This again had translating wing roots, and for various reasons was unsuccessful. But by this time a team in Britain, led by Sir Barnes Wallis at the Vickers-Armstrongs works at Weybridge, had come up with a better formula. The key factor was that the swinging wings were pivoted at the outer (rear) extremities of a large fixed wing root, called a glove, with a sharply swept leading edge. This enabled the pivots to be fixed.

As related earlier, Britain did nothing with the idea, though Wallis touted it in various forms, most having the family name of Swallow and being distinguished by engines pivoted well outboard along the wings, or even at the tips. Wallis had been so shocked by the war that he had become a pacifist, and his Swallows were supposedly civil. However, in 1957 the British government handed his ideas to the Americans, who grasped them eagerly for military purposes. John Stack, of what a year later became NASA, was the chief architect of what was probably a more practical VG aircraft, a traditional fighter with fuselage-mounted engines but with a Wallis-type wing. In 1962 this scheme formed the basis of the submissions in the USAF Tactical Fighter Experimental (TFX) submissions.

The story of how the TFX requirement became the F-111 has been told many times. At the start the programme was expected to lead to a fighter that would not only replace virtually all the fighters and attack aircraft in the USAF, but, in the absence of the slightest competition from the British, would sell throughout the world. Among its advanced features were, of course, VG wings, as well as augmented turbofan engines, the possibility of thrust reversers, titanium primary structure, powerful multimode radar and other advanced equipment, and all-round performance transcending anything seen previously. For example, the original requirement, called SOR-183 (Specific Operational Requirement), asked for Mach 2.5 at height and 1.2 at sea level (flying under the control of a new invention, terrain-following radar), the ability to operate from a strip 3,000ft long bulldozed from the jungle, and an unrefuelled ferry range of 3,300 n.m. (3,800 miles). These demands proved impossible to meet, and they were compounded in February 1961. After three weeks as Secretary of Defense, Robert S. McNamara studied the requirements for TFX, and for the Navy's Fleet Air Defense Fighter, and decided that roughly a billion dollars could be saved by meeting them with two versions of the same aircraft. When both services disagreed, he took the unprecedented step, on 1 September 1961, of not only giving an explicit order launching a bi-service TFX, but even of telling both parties *how to design the aircraft*, writing down maximum lengths, maximum weights, minimum internal fuel, radar antenna sizes, weapon loads, and much more. His attitude was, 'Those designers just aren't trying hard enough!'

The two finalists were Boeing and General Dynamics (GD). So important was this programme that, uniquely, the proposals were refined in four rounds of evaluations. In the final round every member of the Systems Source Selection Board, comprising representatives from Tactical Air Command, Systems Command and Logistics Command, the Navy Bureau of Weapons, and other user bodies, unanimously voted for the Boeing submission, which among other things was superior in having CCV unstable design (Chapter 11), titanium structure, reversers (usable in flight), highly efficient dorsal inlets, greater bomb load, better manoeuvrability, longer ferry range, and many other advantages. It was therefore almost unbelievable when, ignoring a year's evaluations by hundreds of expert people, McNamara awarded the contract to the rival, GD. And General Electric never understood why their outstanding MF295 engine was simply dismissed as 'unacceptable'.

So GD built the F-111A, powered by Pratt & Whitney TF30 engines, while Grumman was a partner on the carrier-based F-111B for the Navy. The wing was mounted high, the quarter-round inlets being tucked beneath the roots. Other features included huge tailerons giving control in pitch and roll, two enormous mainwheels giving a spongy touchdown yet folding into the fuselage, a giant door airbrake making it impossible to carry bombs or tanks under the fuselage, side-by-side seats for pilot and right-seater, and a remarkable crew-escape capsule reminiscent of that devised by GD for the B-58. Side-by-side seating seemed odd for a Mach 2.5 fighter, especially as each crew-member had no external view except ahead and to one side.

GD flew the first YF-111A on 21 December 1964. There followed perhaps the most troubled flight-test programme in history, characterised by excessive growth in weight, escalation in cost, drag much greater than prediction and a propulsion system so worrying that it was officially called 'a hazard to safe flight'. Years later the various F-111 production versions were still suffering from these troubles, as well as from catastrophic failures of primary structure, the flight-control power units and even the crew-escape system. The F-111B was soon abandoned. Altogether, even including a sale to Australia (of 24 aircraft which cost three times the original quoted price, and were ten years late), only 562 F-111s were built instead of the planned thousands. All of which overlooks the fact that the F-111 was not a fighter at all, but a formidable bomber, which from Vietnam to Libya and Iraq flew long missions of the most demanding type.

When the US Navy finally gave up on the F-111B, it held a quick VFX fighter competition, and picked Grumman on 15 January 1969. Remarkably, the Long Island company flew the first F-14A Tomcat as early as 21 December 1970. Despite having engines similar to those of the F-111, so that for the first 15 years there were severe problems, the F-14 proved to be a superb aircraft, unique in the Western world in having an internal gun and three different species of air-to-air missile, including the 100-mile-range Phoenix which was carried by no other aircraft. Wing sweep and tailplane geometry were similar to those of the F-111, but after much study it was decided to position the tailplanes below the wing, and to fit twin vertical tails, canted outwards, giving good stability and control to an AOA of almost 90°, with minimal aircraft height and length for carrier compatibility. To preserve CP shift, and reduce wing loading, small glove vanes could pivot out from the leading edge as the wings hinged back to the maximum of 68°. The engines were spaced well apart to give room for extra internal fuel, the total being a useful 1,986 gallons. With such endurance, and two engines and two crew, the F-14 had enviable capabilities. Avionics were progressively upgraded, and from 1984 Grumman's effort was devoted to improved versions powered by the GE F110 engine.

In Britain the vacuum left by the cancellation of all major programmes led to an attempt to produce an Anglo-French variable-geometry (AFVG) fighter. The French walked out, but in July 1968, a Memo of Understanding was signed by Britain, Germany, Italy and (briefly) the Netherlands concerning an MRCA (Multi-Role Combat Aircraft). The first MRCA flew on 14 August 1974, and as the Panavia Tornado entered service in 1980. By 1998 a total of 992 had been delivered. All have shoulder-high wings with advanced high-lift systems pivoting from 25° to 67°, a smaller arc than most VG aircraft. As on the F-14, the tailerons are significantly lower than the wings, but there is a giant single fin and rudder. Other features include tandem seats, reversers, and the ability to carry an 8,000lb bomb load while leaving the wing pylons free for tanks, EW payloads and AAMs. The air-defence variant (ADV), designated Tornado F.3 by the RAF, has a more pointed radome over a different radar, and a fuselage lengthened to carry tandem Sparrow or Sky Flash missiles recessed underneath. This extra length increases internal fuel by 250 gallons. All versions have the exceptional ability to reach over 800kt (920mph, Mach 1.21) at sea level, of

course in the clean condition. The VG wings make a difference in low-level full-throttle ride quality that has to be experienced to be believed. Aircraft with fixed wings cannot achieve such sea-level attack speeds and figuratively 'shake the pilot's eyeballs out'.

When France pulled out of the AFVG programme, Dassault was deeply engaged in the Mirage G programme, a range of VG fighters with either one TF306 engine or two Atars. These got nowhere, nor did the fixed-wing *Avion de Combat Futur* (ACF), and eventually the choice fell on the much smaller Mirage

The very first Tornado, flown in 1974, was aerodynamically little different from the 991-odd that have followed it. When this photo was taken the wing pivots had not been faired.

2000, looking from a distance like a Mirage III. But in the Soviet Union they take decisions not on the basis of fashion but on numerical results, and VG wings swept the board. Some were conversions of established types. The most numerous were a family of tactical attack aircraft developed by the bureau named for P.O. Sukhoi. Throughout the late 1950s, when Sukhoi was still in charge, this bureau spent more time than any other studying nose inlets for supersonic aircraft, especially types fitted with radar. The aircraft actually flown were large, powered by the afterburning turbojets in the 15,000 to 24,000lb class by the Lyul'ka bureau. The S series, with a 62° swept wing, led to the Su-7 family of attack fighters, while the T series, with 57° delta wing, led to the Su-9 and Su-11 interceptors. These in turn led to the twin-engined Su-15 family (Chapter 11), one of which was a STOL research aircraft with extended wings and three lift jets in the centre of the fuselage. Meanwhile, an Su-7 was rebuilt in 1965–6 with VG. outer wings. Designated Su-7IG (*Izmenyaemya Gayometriya*), this aircraft was little changed apart from structural redesign of the inner wing, the addition of huge fences at the junction with the outer panel, and the fitting of outer panels with full-length slats, the pivots being at about 50 per cent semi-span, and offering sweep angles from 28° to 62°. The resulting Su-17 family have a combat radius with a 5-tonne weapon load 30 per cent greater than the Su-7 with 2 tonnes. Over 1,500 were built.

The Central Aero- and Hydrodynamic Institute also assisted both the A.I. Mikoyan and P.O. Sukhoi design teams with VG fighter and attack aircraft of completely new and uncompromised designs. The first

to fly was the Mikoyan 23-11/1, demonstrated at the Aviation Day airshow on 9 July 1967. Accompanying it was another new MiG, the 23PD. The latter was a STOL tailed delta, in some ways broadly resembling a MiG-21 but considerably bigger, and with the wing in the mid-position. The bureau had previously flown a MiG-21 PF rebuilt with two lift jets in the centre of the fuselage, and the 23PD incorporated almost the same lift-jet installation. The bureau would have liked this to have been accepted, because it minimised development and appeared simple. No attempt was made to fly so slowly that reaction control valves were needed at the extremities, as in STOVL aircraft. Minimum speed was in the order of 50kts (58 mph, 93km/h), but this was still low enough for operations from a 200m (656ft) strip. Nevertheless, the VVS (air force) wanted to compare this with what might be done with a VG wing, and the result was the 23-11/1. This naturally had a totally new wing, mounted high on the fuselage, and aligned with the tailerons as in the F-111. At maximum sweep of 72° the outer panels were aligned with the fixed inboard section. At minimum sweep of 16° the leading-edge droop flaps and all three sections of trailing-edge slotted flap could be used, giving excellent lift and control down to 220km/h (137mph). Under the wing were the inlet ducts, fed by F-4 type lateral inlets. The new wing meant that the main landing gears had to fold into the fuselage, and another innovation for a MiG was the addition of a large sideways-folding ventral fin.

The production choice fell on the VG aircraft, which entered service as the MiG-23 all-weather interceptor, with secondary attack capability. In production form this had a different engine, the Lyul'ka turbojet being replaced by the Tumanskii R-27 rated with afterburner at 22,485lb, and considerably simpler and lighter. Later this was supplanted by the Tumanskii R-29B, rated at 27,500lb. There were many other refinements, including

When cleaned up, the Yak-41M was highly supersonic.

A MiG-27M of the Indian Air Force. Over 4,500 of this swing-wing family were built.

the addition of giant claw-like dogtooth projections at the leading-edge root of the outer wings. At minimum sweep these meet the leading edge of the fixed inboard wing, but in supersonic flight they project ahead, at either of the selectable sweep angles of 45° or 72°, so that in hard manoeuvres at up to 8.5 g they cause a high-energy vortex across the top of the wing. From the MiG-23 was developed a series of dedicated attack aircraft, all having a downsloping 'ducknose' with the radar replaced by a comprehensive group of attack sensors and devices, and with heavy protective cockpit armour. The weapon pylons, which include two on the flanks of the rear fuselage, are tailored to bombs and other attack weapons, and the main gears have larger low-pressure tyres (causing visible bulges in the rear fuselage) for operations from unpaved strips. The MiG-27 is a family of even more dedicated attack aircraft, with a simpler propulsion system comprising an R-29-300 engine fed from fixed inlets, and with a short two-position afterburner nozzle, maximum thrust being 25,350lb. This installation deliberately ignores high-Mach considerations, and is tailored to subsonic cruise at sea level, though in clean condition Mach 1.1 can still be reached at treetop height, and 1.7 at altitude.

The partner to these attack versions is the Su-24. Considerably larger and about twice as heavy as the MiG-23/27 family, this was designed as an almost exact answer to the TSR.2. Like the British aircraft, the T-6 series of prototypes were given a delta wing of considerable area (680sq ft), with almost identical blown flaps and down-turned wing tips. The first prototype, the T-61, was said to have had 'additional engines', and certainly a battery of eight of the standard RD-36-35 lift jets were intended to be fitted to one of the fixed-wing T-6 prototypes, but T-61 itself is now on view at Monino and it has no lift engines. In 1967 it was evaluated against the T-62, with a VG wing similar to that of the Mikoyan 23-11/1, though larger. The pilot can select settings of 16°, 45° and 68°. Unlike Mikoyan, the Sukhoi bureau picked leading-edge slats and double-slotted flaps. Because of the size of the flat-plate scanner of the main radar, it was decided to seat the pilot and weapon-system officer side by side, as in the

Recent photographs of the enormous Tupolev Tu-160 'Blackjack' bomber. *Aviatsiya-Kosmonautika*

F-111. Despite the fact that this is a sea-level strike and reconnaissance aircraft, it was decided to fit fully variable inlets and full-length con/di nozzles, giving Mach limits of 1.15 at sea level and 2.18 at high altitude. Six versions are in use, the latest being dedicated to reconnaissance and EW (electronic warfare). With a maximum weight close to 90,000lb, and mission radius with a 6,614lb bomb or missile load exceeding 800 miles, these are formidable aircraft, and over 800 were delivered.

In contrast, production of the enormous Tu-160, discussed later, is almost measured on the fingers. This strikingly beautiful strategic bomber is virtually a USAF B-1 on a larger scale. In turn, the B-1 resulted from the

longest period of study ever devoted to any aircraft, as hinted in Chapter 8. One of the most intractable problems facing aircraft designers is that modern warplanes tend to need about eight years between conception and first flight, and another eight to get into service in numbers, by which time the concept is outmoded, the world scene has changed, and everyone can see what totally different aircraft should have been designed. This explains why so many of the aircraft in Chapter 8 never went into service, a classic example being the B-70.

To replace it, the USAF agonised over a succession of manned delivery systems, at the same time wondering whether the recently created ICBM force really did make such aircraft superfluous. Some had hardly any wing, and were designed for the highest possible supersonic speed at low level. These bullet-like aircraft were non-starters, partly on account of the sheer difficulty of flying so fast near the ground, except over sea or flat desert, partly because even a virtual absence of wing still gave the crew a very rough ride, and above all because there was no way enough fuel could be carried for a useful combat radius. Nuclear energy was again studied, and again found to be impossibly difficult. In the end the choice fell on a conventional aircraft, but with VG wings and extremely careful overall design. Requests for proposals were issued in November 1969, and after evaluation of the bids by Boeing and North American Rockwell, the latter's submission was picked in June 1970. The first of four B-1A prototypes, painted white, made its first flight on 23 December 1974.

In size, the B-1A was smaller than a B-52 or SST, but was still very impressive. At minimum sweep of 15° the span was almost 137ft, and the fuselage length was 10ft greater. At maximum sweep of 67.5° the span was reduced to just over 78ft. Each outer wing was fitted with a full-span slat, and seven segments of single-slotted flap. Roll control was provided by upper-surface spoilers, and by differential operation of the tailplanes, mounted halfway up the fin. A prominent feature was that each outer wing was pivoted to an inner section blended into the shapely fuselage. The diffusion-bonded titanium wing carry-through box showed no point at which one could say the fuselage had ended and the wing begun. These large volumes along each side of the aircraft were from the start to be filled with electronic defence systems, with numerous flush or suppressed antennas. Under the inner wings, as far aft as possible, were the two nacelles each housing twin General Electric F101

turbofans, each rated at 30,780lb with afterburner. As befits an aircraft capable of Mach 2.22 at high altitude, the inlets were fully variable, bearing some resemblance to the bigger inlets of the XB-70. The bogie main gears were just inboard of the nacelles, folding inwards into the fuselage. Again like the XB-70, the crew numbered four, housed in an ejectable survival capsule. Unlike the XB-70, there were three 15ft weapon bays in a row, each able to carry a rotary launcher for eight AGM-69A (SRAM) missiles or 25,000lb of bombs. Another new feature was a low-altitude ride control (LARC), later restyled the structural-mode control system (SMCS), which sensed vertical accelerations in turbulence and countered them by automatically driving delta foreplanes sloping down on each side of the nose. From the start, the B-1 enjoyed top priority whenever the USAF fought for funds, but it was nevertheless vulnerable. President Nixon saw it as a bargaining counter in the SALT talks. President Ford liked it so much he got his Defense Secretary to fly it. President Carter disliked it so much he cancelled it, saying the old B-52 could soldier on with a supposed 'new weapon', the cruise missile. And President Reagan resurrected it, ordered 100, and saw them all put into SAC service.

The aircraft the USAF finally got is the B-1B Lancer. Apart from having the white paint replaced by what looks like dark charcoal, but is actually two shades of grey and a deep green, the production bomber differs from the B-1A far more than it appears to. The basic change is that Mach 2.2 has been forgotten, and the B-1B is designed for Mach 0.99 (652kts, 1,207km/h) at 500ft. This is reflected mainly in the simpler fixed-geometry inlets specially configured to minimise RCS (radar cross-section). By the time the B-1B was designed, in 1980–83, RCS had become far more important than speed, and everything possible was done to reduce it, though not much was done to reduce visual, aural and infra-red signatures. At the same time, the defensive avionics, built around the vast ALQ-161 system, were greatly upgraded and advanced in capability. Other changes included replacement of the crew capsule by conventional ejection seats, a great increase in internal fuel capacity (raising maximum weight from 389,800lb to 477,000lb, and requiring structural redesign of the wings and main landing gear), redesign of the internal weapon bays with a movable bulkhead to permit installation of an eight-shot rotary launcher for the AGM-86B cruise missile (too long to fit in the original

The Rockwell B-1B Lancer. Though variable in profile and area, the F101 engine nozzles do not have to cater for over Mach 1. *PRM Aviation*

bays), and the addition of six racks under the sides of the fuselage each able to carry two more ALCMs. Maximum weapon load is 75,000lb internal, and 59,000lb external.

The first B-1B flew on 18 October 1984, and all 100 were delivered by 30 April 1988. Perhaps predictably, this exceedingly complex and costly ($282m) aircraft became the focal point of an amazing nationwide storm of abuse and criticism. Like the Boeing SST and Concorde

before it, this criticism was an interesting psychological phenomenon, totally unrelated to any good or bad features of the aircraft itself. The media latched on to the slightest suggestion of any problem, and fanned it into huge proportions, while totally ignoring any impressive achievements, and never reporting any of the detailed briefings given in Washington or by the crews of the Bomb Wings that proudly flew the B-1B. I do not pretend to understand the reasons for this attitude, which saddened the dedicated personnel of SAC, who know the aircraft better than the shrill knockers.

I am assured that this kind of mindless abuse is not seen in the Soviet Union, directed against the Tu-160. As noted earlier, this is basically a bigger edition of the

A contrast in supersonic wing design is afforded by this
B-1B escorted by an F-106 chase plane.

The Soviet Union believed the mighty Tu-160 was a better aircraft than the B-1B. It has nearly double the engine power, and the author accepts the claim that it has a smaller radar cross-section, despite its greater size. *PRM Aviation*

B-1B. Compared with the US bomber the Tu-160 has retained all of its high-altitude performance, with a peak Mach of 1.92. The twelfth aircraft, inspected by a US delegation at Kubinka on 2 August 1988, was painted white, though most are pale grey. Western observers commented on the unbelievably smooth exterior skin, without a trace of a joint or blemish. Though this improves flight performance, the main reason is to minimise RCS. On the right side of both pilots are wing levers for selecting sweep from 20° to 65°. At minimum sweep the span is 182ft 9in, and fuselage length is 177ft. Gross weight is 606,260lb, at which a 36,000lb load of bombs or various cruise missiles can be carried over a radius of 4,535 miles, about 1,000 miles greater than for the B-1B. The crew of four board via a powered equipment platform and ventral hatch, thence walking along a corridor between the main avionics bays. Other differences from the B-1B are that there is no separate rudder, the upper fin (above the tailerons) being pivoted as a single unit, and that the main gears each have three pairs of wheels.

Col Evgeni Vlasov recently published a critique comparing the Tu-160 with the B-1B, which left one thinking the USAF bomber must be an also-ran. He emphasised the Soviet giant's superior aerodynamics, much lower radar cross-section, and generally better stealth qualities. A detail is that, by putting the crew compartment ahead of the nose gear, the body cross-section was significantly reduced. The Tu-160 needs no external weapon attachments, nor has any AOA or other handling restrictions. Each NK-321 engine is rated at 55,115lb, compared with 30,000lb for the B-1B. Not least, Vlasov insists that, though 'more than 100 computers operate aboard the aircraft', everything works, including the entire electronic-warfare system – unlike the B-1B, he says. By 2008 the number of active Tu-160s had dwindled to 16, though these are gradually being put through a comprehensive upgrade. The first to be returned to service flew to Engels airbase on 29 April 2008.

Other supersonic aircraft of today are discussed in the next chapter.

Chapter 11

TODAY'S SUPERSONIC AIRCRAFT

For nearly 40 years, the Boeing (originally McDonnell) F-15 Eagle has replaced its forerunner, the F-4 Phantom, as king of Western fighters, and South Korea bought a new-build batch in 2008. It won the USAF FX competition in 1969, the need for a new air-superiority fighter having been spurred by the MiG-25. No attempt was made to match the Soviet aircraft's speed, the design clean Mach number being 2.5. The decision was taken to use a high-mounted wing of 608sq ft area, with no variable geometry apart from ailerons and plain flaps. Under the wing pass the ducts, joining the lateral horizontal-2D inlets, with variable upper walls, to the augmented turbofan engines mounted close together. The original engine was the P&W F100 rated at 23,810lb, but GE succeeded in giving customers the option of the F110 rated at up to 32,130lb. This engine arrangement leaves no space between them for fuel, and the twin vertical tails are mounted on beams cantilevered back from the wing, which also carry the slab tailplanes. The latter have a giant dogtooth discontinuity on the leading edge. In the fairing aft of the canopy, above the wing, is a large speed brake, used after touchdown plus the drag of the aircraft held in a nose-high attitude. After lowering the nosewheel to the ground, the carbon brakes on the main wheels are used, there being no reversers or parachute. Medium-range radar-guided and close-range IR (Sidewinder) missiles can be carried under the fuselage and wings, and an M61 gun in the right wing is fed from a drum in the fuselage, the belt passing over the right engine air duct.

There are no canards, forebody strakes, fences, droops, slats, dogteeth (except on the tail) or ventral fins. In the author's view the only obvious shortcoming is that, to rely on a combat aircraft which has tyres inflated to 340lb/sq in, for paved runways, and gets airborne at 240mph, meaning quite a long runway, shows a total lack of imagination of what would happen in any future war against a major power. Today the production version is the two-seat F-15E multi-role attack fighter, weighing up to 81,000lb. This is clearly an even better weapon

A total rebuild of the first two-seat F-15B, the S/MTD had 2D vectoring nozzles and powered canards derived from F/A-18 tailplanes.

This F-15C served with the 48th Tactical Fighter Wing at RAF Lakenheath, England. *PRM Aviation*

for use against primitive enemies that cannot hit back. A small step in the right direction was the F-15S/MTD (STOL and manoeuvring technology demonstrator), with canards and 2D vectoring nozzles on the engines. These changes made possible new kinds of agility in flight, but in war their value would be in enabling an S/MTD aircraft to operate from a 1,500ft runway. Of course, even that would be hard to come by, but at least the USAF is beginning to recognise that airbases are the first things that would disappear in a real war.

The canards of the S/MTD were adapted from the tailplanes of an F/A-18 Hornet. This aircraft was developed from the Northrop YF-17, one of two lightweight fighter

(LWF) contenders ordered by the USAF in 1972. The other was the General Dynamics YF-16. Both were in the 22,000lb class, with thin wings tapered on the leading edge, and carrying tip-mounted Sidewinders. Whereas the YF-17 had two GE YJ101 engines, the YF-16 had an F100, virtually identical to that fitted to the twin-engined

A dramatic photograph of an F-15E taking off with full afterburner. *PRM Aviation*

This F-16C served with one of the USAF Aggressor units, either the 16WS, 64AS or 414CTS, painted to look like a 'bad guy'. *PRM Aviation*

F-15. This commonality was a big plus, as was the greater combat radius gained by the F100's lower fuel burn. Both aircraft had an accent on simplicity, with fixed inlets, though both had multimode radar. The YF-16 was chosen, and, developed into the F-16 Fighting Falcon, has been procured in numbers exceeding any other Western fighter except the F-4 (which it will overtake) and F-86. The USAF were reluctant to take an LWF seriously, thinking it must be 'inferior' (to the F-15), but agreed to buy some when a potentially large market opened up in Europe. The initial 650 later jumped first to 1,388 and subsequently to 2,729, while in 2008 some 1,790 more had been ordered by 22 other air forces. Features include a blended wing/ fuselage design, single fin, tailerons supplementing small flaperons (ailerons which droop at high AOA), fly-by-wire flight controls electrically signalled from a small force-sensing stick on the right side of the cockpit, and a backward inclined seat mounted high inside a totally transparent polycarbonate canopy giving virtually perfect all-round view. The F-16 is an RSS (CCV) aircraft, as described later.

The rival YF-17 was not abandoned, but used by McDonnell Douglas (now Boeing) as the basis for the F/A-18 Hornet for the Navy. Compared with the

YF-17, this had more powerful F404 engines of 16,000lb each, much greater internal fuel capacity, APG-65 radar, carrier equipment, and the ability to carry heavy attack loads or radar-guided missiles. The F/A-18 designation underscores its versatility, for it was designed to replace the F-4 in the fighter role and the A-7 in the attack role. At first it suffered from poor combat radius, inability to meet wind-over-deck speeds and poor rate of roll. These were cured, and the aircraft's good features soon made it popular, and even brought sales to countries with no aircraft carriers (nobody bought the Northrop F-18L, with carrier features removed). Like other modern warplanes, the F/A-18 has a traditional unswept wing, but twin vertical tails, canted out at 20° for control at high AOA. Deliveries to the Navy and Marine Corps began in May 1980. Unexpectedly rapid fatigue damage from high-AOA flight was suffered in service, resulting in an urgent aerodynamic rethink. The basic problem was that the inner wings are continued forwards, in a sharp-edged leading-edge extension (LEX). This plays a major role, creating vortices, which at AOA up to about 20° coil harmlessly past the fins. Reduced pressure above them provided a convenient place to suck out the inlet boundary layer, through a large rectangular slot on each side. At higher AOA the vortices impinged on one or both of the fins, causing cracking of the fin attachments and rear fuselage structure. The back end was redesigned to resist severe fin loads, and the vortex strength and direction was changed by adding a fence on each side, about halfway between the LEX/fuselage junction and the root of the outer wing. Except for test and research flying, high-AOA flight was forbidden, though the F/A-18 designers hoped eventually to make such flight permissible at air displays, if not a routine part of the Hornet pilot's combat repertoire, to rival the Russians.

In 1991 an enlarged and upgraded Super Hornet was proposed, initially to replace the unbuilt Grumman A-12 'stealthy' attack bomber. The result is the F/A-18E, two-seat F/A-18F, and the E/A-18G Growler, the replacement for the EA-6B Prowler electronic-warfare platform. Powered by GE F414 engines rated at 22,000lb each, these new versions have a further major increase in fuel capacity, avionics and weapon options. The first Super began testing on 26 March 1997, and service entry, with VFA-122, was on 17 November 1999. By 2008 orders were close to 500, following 1,478 Hornets.

Having discussed the two LWFs I can no longer avoid relaxed static stability (RSS) and control-configured vehicle (CCV) design. These have been made possible because of another acronym, active controls technology (ACT), which in turn results from dramatic advances in avionics. ACT means that the flight-control system, or a parallel system driving other aerodynamic surfaces, can be left by itself to modify the flight of the aircraft in a favourable way. The first ACT was developed on a Lancaster bomber in 1948–50, for use on the Brabazon. The system sensed vertical acceleration due to gusts (up/down air movement), and drove the ailerons in unison, both up or both down, to cancel out the 'bump'. This never went into service, and it took another 31 years before a system identical in principle was put into use on the extended-span Lockheed L-1011-500 TriStar. Another ACT system is the ride-control fitted to the B-1B (Chapter 10).

A fly-by-wire system transmits the input demands to the surface power unit at the speed of light, and the associated electronic sensor or computer also operates with lightning speed. It follows that today's aircraft designer can, from the very start of design, plan an aircraft in such a way that it relies totally upon a lightning-fast and completely reliable control system in order to fly, to manoeuvre, or to avoid structural failure. This is still a fairly new capability; until about 1970 it was impossible.

Ride control and gust-alleviation are but two of the ancestors of today's fighter flight-control systems. A more direct parent was the yaw damper, of 1949, to stop uncommanded 'snaking', which in turn led by about 1957 to full stability-augmentation systems (SAS). These improve handling, especially in extremes of the flight envelope. Indeed, the flight envelope in many fighters has been appreciably opened out by an SAS, making possible otherwise unattainable manoeuvres. Conversely, an ACT can equally well be used to limit the attainable flight envelope, so that the pilot can never command a high-speed stall and flick, structural failure or any other dangerous situation. Among other acronyms is Spils (stall-prevention and incidence-limiting system), one of which is fitted to the Tornado F.3 interceptor. Today we can design so that AOA need not be limited, and the Russian designers showed the way to making deliberate manoeuvres with AOR equalling, or even exceeding, 90°.

The crucial factor affecting the design of modern combat aircraft was the appreciation some 40 years ago that the development of avionics had made possible a completely new type of flying machine which, being

The F/A-18 Super Hornet has updated engines and improved avionics and weapons options. *PRM Aviation*

The F-4 Phantom could soon be overtaken by the F-16 in the numbers procured by the world's air forces. *PRM Aviation*

inherently unstable, could not be flown manually at all. The idea of flight stability is obvious: a dart or arrow is stable, whereas a dart or arrow travelling 'tail first', the point facing backwards, is unstable. With an unstable body, the slightest disturbance will cause instant departure, the name given to an uncommanded rotation about one or more axes which results in a total change in the body's attitude. One can get an idea of the situation by leaning out of a fast car or train holding a sheet of cardboard edge-on to the airflow. If you hold it by the leading edge, so that it trails downstream, it is stable; it may flutter, but will exhibit no tendency to a major change in attitude. Now try holding it by the trailing edge, so that it projects upstream. If you are incredibly lucky you may hit exactly the right situation at which AOA is zero. There is no force perpendicular to the surface, and the sheet maintains its attitude. But the slightest up or down movement, such as in the real world would occur many times per second because of air turbulence or movement of your hands, and the sheet would depart instantly. During the first fraction of a second, the AOA would increase, resulting in greater up or down lift force, accelerating the rotation and speeding up the process, like the chain reaction in a nuclear weapon. Well within a second the sheet would have reversed direction, probably being torn from your hands. You have no chance of avoiding the process, nor of stopping it once it has started.

One of the many versions of the Mirage 2000 is the 2000N two-seat attack aircraft of the Armée de l'Air. *PRM Aviation*

This is exactly what would happen if a pilot tried to fly an RSS or CCV fighter manually. The process has also been likened to sitting on the bonnet of a 100mph car pushing a bicycle ahead by the handlebars, the bicycle facing in the other direction. This is obviously as impossible as trying to control the sheet of cardboard, but to an electronic control system it would be no problem. Such systems typically check on what is happening 40 times per second, and take any necessary corrective action. If that is not frequent enough, the designer can make it 50 or 100 times per second.

Why do we want to make our aircraft unstable? There are really two reasons. One is that an unstable aircraft is all the time trying to depart from straight and level flight, and has to be continuously held in check, so if we actually command a manoeuvre it will respond more immediately and more positively than a naturally stable aeroplane, which hates to be forced out of straight and level flight. The other is that the RSS or CCV fighter can be made significantly smaller and lighter than a stable aeroplane designed to fly the same mission. This results from the fundamental design of the aircraft.

Conventional aeroplanes – the stable ones, representing over 99 per cent of those in existence today – have their

The 92nd JAS39A Gripen is seen here with all eight pylons empty. *PRM Aviation*

CG significantly in front of the CP of the wings. Some elementary textbooks show a 'balance of four forces', the nose-down couple of one pair (such as lift and weight) being balanced by the nose-up couple of the other (thrust and drag). In practice, this almost never happens, and the pitch equilibrium is provided by a download on the tailplane. This download is typically 10 to 15 per cent of the weight. In other words, almost all aeroplanes behave as if they were at least 10 per cent heavier than they actually are. Thus, they need a bigger wing, causing greater drag, and more engine power, burning more fuel, which completes the vicious circle by adding further to the bulk and weight. They even confer a further self-inflicted injury by needing a large extra download on the horizontal tail at the very times (rotation on take-off, and the landing flare) when lift needs to be a maximum.

The latter fault can be eliminated by using a foreplane, which instead of adding to the weight at the back adds to the lift at the front. But the RSS or CCV concept does not depend on using a foreplane. To explain the terms, relaxed static stability means that the static stability in pitch is reduced, or even made negative, by moving the CG to the rear or the CP to the front. Such an aeroplane could not fly without a flight-control system able to react so fast that, every time instability caused the aircraft to try to depart from its desired trajectory, the control surfaces – tailplanes or foreplanes – were instantly driven just far enough to arrest the tendency. Thus, the entire design of the aeroplane would be driven by the flight-control system; hence the generic designation control-configured vehicle.

I do not want to get too technical, but changing over from natural to artificial stability completely alters many of the fundamental characteristics of the design. In almost every case, the change is for the better. For example, the tailplane of a conventional fighter is always sized by its need to provide static longitudinal stability. If this demand is eliminated, and replaced by fast-acting responses of the powered surface, the tailplane size is determined by its remaining functions of pitch control and pitch trim. Thus, it can be made significantly smaller, resulting in reduced drag, weight and cost. Even more important, the only download on the tail is needed for rotation on take-off and landing (and this can be replaced by a lifting force, if a foreplane is used instead), so the wing can also be made significantly smaller. Thus, overall drag and empty weight can be reduced by 15–20 percent, trim drag reduced by 50–70 per cent, and turn rates increased by 10 per cent (subsonic) to over 20 per cent (supersonic). It all rests on fast-reacting flight controls with essentially perfect reliability. In blind-landing development the target was one failure in each 10 million flying hours, and this has been continued with today's supersonic flight-control systems.

Of course, we think of RSS and CCV design as applying solely to fighters. In fact there is every reason why such techniques should also be applied to transport aircraft, including subsonic ones, but this will take time to become the accepted norm. Already we have seen interesting configurations on such aircraft as the Starship and Avanti, but they were not supersonic. Designers have felt they needed to have the back-up of a lot of full-scale free-flight research before taking decisions on CCV aircraft. Examples have included the MBB F-104CCV, the McDonnell F-4CCV, the Mitsubishi T-2CCV, the British Aerospace ACT Jaguar, the General Dynamics AFTI/F-16, and two NASA programmes, the ACTIVE F-15B and the AAW F/A-18A. The F-16 was already a pioneer CCV design, the original YF-16 having been designed to fly at negative static margins up to 13 per cent. It will be appreciated that an unstable CCV aircraft has to rely totally on its FBW control linkage.

No mechanical reversionary connections are possible, because they could not move fast enough, and the F-16 was probably the first production aircraft since Concorde to dispense with a mechanical back-up system.

The Advanced Fighter Technology Integration (AFTI) F-16 was particularly interesting in that it combined CCV design with a radical flight-control system enabling it to do things that previously were impossible. All normal aircraft, including F-16s, have to fly traditional 'coupled' manoeuvres. For example, to fire a gun against a target on the port (left) bow, the F-16 pilot has to roll to the left, pull the stick back and (probably) press the left pedal, and then eventually roll out behind the target. The AFTI pilot merely applies 'a bootful of left rudder' to swing the nose on target, opening fire in a fraction of a second. He can keep the target 'on the pipper' without the need for any bank, and probably with no change in flight path, merely in the pointing of the fuselage. Of course, you must have a very strong fin and rudder. The AFTI, which flew in July 1982, had such remarkable capabilities that it is surprising they have not been copied. By flying with the small sidestick, the pilot could do anything an ordinary F-16 can do, which is not bad for a start. But by using the stick, pedals and a throttle twist-grip in a new way, the AFTI pilot could perform six decoupled, or six-degrees-of-freedom, modes that in a combat mission would multiply the aircraft's capability.

Three of these are in the horizontal plane. Direct side force uses the downsloping canards, rudder and roll-control surfaces to control lateral acceleration with zero sideslip. In other words, the aircraft can be made to fly in a curved path to left or right, very quickly, with no need to bank the wings; over 1.5 g can be pulled laterally. Yaw-axis pointing enables the yaw axis, the way the fuselage is pointing, to be varied over considerable angles in a fraction of a second while maintaining the original flight path, and thus with zero lateral acceleration. Lateral translation enables the pilot to control lateral velocity at a constant aircraft heading. Thus, while continuing to point in the same direction, the aircraft can instantly be made to crab sideways, in what could be called 'a square turn', and thus follow a different track across the ground. The value of these capabilities in air- to-air or air-to-ground firing, bomb delivery, cross-wind landing, and many other situations is obvious.

The other three types of decoupled motion are in the vertical plane. Direct lift provides control of normal

(perpendicular to the longitudinal axis) acceleration at constant AOA. This uses the wing flaperons, interconnected with the horizontal tails, to give precise vertical flight-path control, an increased load factor at unchanged AOA, and quicker dive recovery. Pitch-axis pointing provides control of pitch attitude without changing the flight path. Again, the flaperons and horizontal tails are used, but in this case to tilt the fuselage through any desired angle, nose-up or nose-down (within limits, of course), while holding the original (e.g. level) flight path. The third longitudinal mode is vertical translation. This controls vertical velocity, at a constant pitch attitude. Without altering the attitude of the aircraft, it can be made to fly up or down. These capabilities add to the previous three modes to create an unbelievably capable vehicle for all forms of weapon delivery, strafing, formation flight, formation aerobatics, or anything else requiring changes in either attitude or trajectory (separately, instead of always together). So far, the only beneficiary of this work seems to be Japan's FS-X, derived from the AFTI and F-16C.

Such considerations of agility, allied with low-observable stealth requirements, have replaced sheer flight performance as the dominant factors driving the

This F-16 served with the 522nd FS 'Fireballs' at Cannon AFB, New Mexico. *PRM Aviation*

design of today's combat aircraft. Indeed, for obvious reasons of noise and infra-red emission it is extremely difficult to see how any stealth aircraft can be made supersonic, and for that reason the F-117A and B-2 do not appear in this book. At the same time, until recently at any rate, there was no discernible wish on the part of the world's air staffs to give up traditional supersonic fighter and attack aircraft, even though they are deafeningly noisy heat-pumping lighthouses.

Perhaps the noisiest and hottest (literally) of today's fighters is the Dassault-Breguet Mirage 2000. Though seemingly almost identical to the mid-1950s Mirage III, this is misleading. The 2000 has major advances, the most important being powerful leading-edge slats, which in tight manoeuvres are opened to give variable wing camber, and hence greater lift, the slat angle being 17.5° downwards inboard and 30° outboard. The variable

This MiG-29 served with the Hungarian Air Force, which is expected to withdraw the last in 2010. *PRM Aviation*

camber is completed by inflight use of the elevons. A further advantage is that, on landing, the elevons can be depressed to increase lift, whereas in earlier delta Mirages they have to be raised, opposing lift. Another plus is a degree of blending of the wing into the fuselage.

Two of the world's best and most important fighters are the Soviet Mikoyan MiG-29 and Sukhoi Su-27. As has often been the case, the Sukhoi is a larger edition of the Mikoyan aircraft, and both rested on basic aerodynamic research carried out by the Central Aero- and Hydrodynamic Institute. Equally, both were disappointing when the prototypes were tested, and only after substantial redesign – especially of the Sukhoi – did they achieve the superb standard of the production aircraft. After all, there's not much future in starting out almost ten years later to create a fighter to defeat the F-15, and coming up with one that was actually inferior! The problems at the Mikoyan bureau were minor, and by the time the prototype MiG-29 flew in 1977 it was a fine aircraft. The aircraft that entered service in 1983 were even better, and since then more has been done, notably in increasing internal fuel and switching to digital avionics, fly-by-wire flight controls, and a 'glass cockpit'.

Like the Su-27, the MiG-29 has a 3 per cent wing with 42° taper on the leading edge, attached to a totally different inboard section with a giant forwards LERX extension, and blended into what could be called the front half of a fuselage in the low/mid position. This body tapers off, though in the Sukhoi it is continued in the form of a slim tube which continues beyond the trailing edge and engine nozzles to house the braking parachute. Completely underneath the wing are the two engine installations, widely separated and with a cavernous space between them which, to meet VVS requirements, can accommodate large cylindrical stores. External fuel, however, is seldom likely to be used; the Su-27 internal capacity of 2,790 gallons gives a range of 2,500 miles without drop tanks! Both aircraft have 2D inlets canted slightly outwards, with variable upper walls and with pressure recovery at high AOA – which in these aircraft can mean over 40° – assisted by the flat wing undersurface upstream. A unique feature of the MiG-29 is that, when an engine is started, hydraulic pressure builds up until it drives the forward upper duct wall downwards, completely sealing off the inlet to avoid foreign-object ingestion. Engine air thereafter enters through a row of large suck-in doors in the top of the wing. On take-off, the main inlets open automatically at about 124mph, though in some inflight manoeuvres the upper doors open also. On landing, the inlets are closed upon nosewheel contact. The MiG-29 engine is the Isotov (A.A. Sarkisov) RD-33 turbofan with a maximum augmented rating of 18, 300 1b. The Su-27 engine is the Lyul'ka (V. Chyepkin) AL-31F, each rated at 27,557lb. This aircraft has inlets similar to those of the MiG-29 but larger, and closed off on the ground by a mesh frame which swings up through 45° from a recess in the floor of the duct, simultaneously opening a row of 12 suck-in doors on the underside.

The aircraft are similar aerodynamically, with full-span leading-edge droop flaps, plain inboard trailing-edge flaps, giant vertical tails mounted on cantilever booms extending aft of the wing, outboard of the engines, and extremely powerful tailerons. There are differences, however, notably in that, while the Mikoyan fighter has tailerons and conventional ailerons operating together in all flight regimes, the Sukhoi has no ailerons but adds large ventral fins mounted vertically under the leading edge of each taileron and extending forwards to the wing trailing edge. Both aircraft have a world-beating kit of radar, infra-red search/track (IRST), laser gun ranger, helmet-mounted sight, lightweight GS-301 30mm gun (149 rounds in the MiG-29, 200 in the Sukhoi) guaranteed to hit with the first round, a tremendous array of external weapons (ten air-to-air missiles of five kinds on the Su-27), various air-to-air or air-to-ground stores on the MiG-29A, and rows of chaff/flare dispensers which on the MiG-29 also serve as overwing fences.

Nothing has been ignored in these aircraft, unlike almost all Western warplanes where the problem is usually the budget. To me, not the least impressive thing about them is their handling. The MiG-29's party-piece of stopping dead and falling away in a tailslide means that any idea of AOA can be forgotten, as can the normal design conditions in the engine inlets. Test pilot John Farley said after flying the MiG-29: 'If they're going to make any more like that, I want to be a MiG test pilot when I grow up.' As for the Su-27, if any reader has yet to see its 'Cobra' manoeuvre, I can only describe it as awesome. The pilot disconnects the autostab and flight-control computers, shuts down the engines, and hauls back on the stick. It is possible to rotate nose-up through 120°, while flying straight and level. Then, stick hard forward and open up the power. The flight path never alters, and if you blink hard you might miss the whole thing.

This Sukhoi Su-27 was put through its paces in a spirited flying display at the Paris Airshow in 2005. *PRM Aviation*

Over 1,300 MiG-29s were delivered by 1996, including exports to 21 countries. Sukhoi exported 208 Su-27s, the chief customer being China, followed (by mid-2008) by 388 of the upgraded Su-30M with AL-35F engines. Both fighters have tandem-seat trainer versions. The MiG-29UB has only a small ranging radar, but otherwise retains weapons capability. The Su-27UB retains full combat capability, including the radar, and has slightly taller fins. Navalised versions of both aircraft have flown trials from dummy carrier decks and from the carrier *Tbilisi* (now named *Admiral Kuznetsov*). The naval MiG-29K has folding wings with bulged tips, strengthened main gears, an arrester hook, a modified IRST ball, and a retractable flight-refuelling probe, and was in 2008 in production for the Indian Navy as the *Baaz* (eagle). The navalised Su-33UB was derived from the STOL T-10-24, and features folding wings, canards, twin nosewheels, and 2D vectoring nozzles, similar to those seen three years later on the S/MTD F-15.

So great were the advances aimed at in these aircraft that they were preceded by exceptional research programmes into advanced titanium alloys, welded primary structures,

RSS/CCV design, quad-channel FBW flight controls and thermal soaking at Mach numbers up to 3. In Sukhoi's case the work involved designing a special research aircraft. The T-4 programme was originally for a long-range interceptor, and eventually funding was provided for two flight articles and one static test specimen. Constructed almost entirely of titanium, the T-4 looked like a smaller XB-70, though the wings did not hinge downwards and a single vertical tail was chosen. Features included four Koptchyenko R-79 afterburning engines in a ventral group, a hinged droop snoot nose, and tandem crew of two. General Vladimir Ilyushin made the first flight on 22 August 1972, and subsequent testing took the Mach number up to 3.3. Much further research led

The carrier-based Rafale M is carrying four Mica air-to-air missles under the wings and training (not fired) missiles on the tips. *PRM Aviation*

to the choice of 2.3 as the Mach limit for the T-10-21, prototype of the Su-27 family, and 2.35 remains the design figure today, that for the MiG-29 being 2.30.

The all-round qualities of the MiG-29, with traditional mechanical controls, and the even more impressive Su-27, have done much to cause Western design teams to rethink their choice of the canard delta as the fashionable shape for a fighter. Indeed, the most advanced fighters in the sky today, the USAF Advanced Tactical Fighters as described later, appear to have the traditional layout.

But in February 1980 Israel Aircraft Industries went ahead with the Lavi (young lion) fighter/attack aircraft, followed in June by the Swedish Saab JAS 39 Gripen (griffon). In 1983 Dassault-Breguet decided to build a prototype Rafale (hurricane), and, following many years of unbuilt projects with similar layouts, British Aerospace went ahead in the same year on the Experimental Aircraft Programme (EAP). All of these aircraft are canard deltas. The Lavi was soon abandoned, but formed the starting point for an impressive Chinese fighter, the CAC J-10, powered by an imported Russian AL-31FN turbofan rated at 27,558lb, much more powerful than the P&W PW1120 of the Lavi. The Gripen and Rafale have supported major production programmes, the former also being in production for South Africa, and the Rafale being in service in several versions, including one for carrier operation, replacing the Crusader. The EAP led to the Eurofighter Typhoon, powered by two

Eurojet EJ200 engines each rated at 20,250lb, of which production was by late 2008 almost halfway through an originally planned total of 638 for the UK, Germany, Spain and Italy, plus 18 for Austria and 72 for Saudi Arabia.

Again, all these aircraft have tandem-seat versions, triplex or quad-redundant FBW flight controls, augmented turbofan engines (one in the Swedish and Chinese aircraft, two in the others), a high proportion of composite construction (mainly carbon fibre and Kevlar), and plain air inlets without variable geometry. The J-10 and Typhoon have ventral inlets, the Rafale has inlets recessed into the lower sides of the forward fuselage, and the Gripen has lateral inlets on the sides. Design Mach number varies from 1.8 to 2.0. The negative static margin, the degree of longitudinal instability, varies from near zero in the Gripen to about 12 per cent for the Rafale and 18 per cent for the Typhoon. The two latter aircraft were designed primarily for the air-combat fighter mission; the standoff-kill, ground attack and reconnaissance missions are secondary, in contrast to the Gripen. Thus, the Typhoon and Rafale are deeply unstable, with thrust/weight ratio up to 1.3 in the fighter mission. It is a common fallacy to think that money can be saved by buying some previous-generation product. Though beloved of politicians, to buy a fighter designed in the 1965–75 era, such as anything from F-14 to F/A-18, no matter how 'upgraded' it might be, would simply be to guarantee to lose every time such a fighter met something designed in the 1980s.

Outside the Soviet Union the most advanced fighters flying in 1990 were the aforementioned ATFs of the US Air Force, the Lockheed (plus Boeing and GD) YF-22 and the Northrop (plus McDonnell Douglas) YF-23. These aircraft represented a completely new generation, perhaps the biggest single leap in combat-aircraft technology since the mid-1930s. Of course, the USAF asked for everything technically possible, with especial emphasis on high reliability, low personnel and maintenance costs, and instant availability for perhaps 40 years.

Among the more obvious features of these prototypes were: stealth design, the first time this has been possible in a fighter; RSS/CCV design for maximum agility; new engines, setting wholly new standards in performance, economy, light weight and maintainability; supercruise capability, meaning the ability to cruise at high supersonic speed without using afterburner; and fully integrated digital avionics with both central and distributed processors in the form of plug-in 6in × 6in cards about 0.5in thick. Two prototypes were built of each fighter, one of each type with the Pratt & Whitney F119 augmented turbofan and one of each type with the General Electric F120 variable-cycle engine (Chapter 7). As noted earlier, the T/W ratio of either engine was roughly 50 per cent greater than anything produced previously, and this played a central role in achieving the desired aircraft agility and all-round performance. Both aircraft were twin-engined, and thus had something like 60,000lb installed thrust, compared with the normal combat gross weight of about 50,000lb.

Both aircraft had almost the same wing, of 43ft span and about 3 per cent thickness, tapered sharply and equally on both leading and trailing edges. Both had outboard ailerons, inboard flaps and full-span leading-edge droop flaps to give variable camber in combat. Other similarities were that both aircraft had a forward fuselage which quickly disappeared aft of the cockpit, twin diagonal fins and conventional tricycle landing gear with a single wheel on each leg. There were, however, significant differences, and it may not be too much of a generalisation to suggest that Lockheed went for performance and agility, with 2D vectoring nozzles, and Northrop went for stealth, without vectoring. Both provided space between the engines. Northrop's separation was greater, despite the fact that for stealth reasons the nozzles were hidden inside large outer casings. Again, the inlets were entirely different, though both were separated more widely than the engines, so that the ducts curved inwards. In the YF-22 the ducts formed the outer part of a broad structure which could be considered either as a wide fuselage or, where the upper surface was concerned, as the centre section of the wing. The forward fuselage had a diamond-shaped cross-section, so the inlets were huge diagonal parallelograms. The sloping inner face was variable in profile, and the lip stood several inches away from the sloping fuselage wall in order to separate off the boundary layer air, which was discharged through the upper surface of the wing. In contrast, the YF-23 had no such inboard section of wing, which maintained its 3 per cent profile all the way to the centreline. But it is not as simple as that. To achieve the desired internal fuel capacity the forward fuselage was considerably longer. More remarkably, the inlets were under the wing, the ducts curving inwards and upwards to the engines mounted above the trailing edge. As in

Typhoon F2 of 29 Squadron RAF. In October 2008 the RAF took delivery of the first of 91 new Typhoon aircraft from the second delivery batch. *PRM Aviation*

the Lockheed, the YF-23 main gears retracted into side bays left by the inward curvature of the ducts.

In both fighters there was a sharp chine, not as pronounced as in the SR-71, starting at the pointed nose and curving up to meet the wing. This served both stealth and aerodynamic purposes, creating strong vortices at high AOA, with plenty of room for them to pass between the tails without causing fatigue. The YF-22

ought to have had better inlet pressure recovery in most flight conditions, and probably lower drag, though the tail of the YF-23 is impressive. It comprised nothing but the two all-moving 'ruddervators', inclined at 40° from the horizontal and abutting against the engine nozzle surround (i.e. rear fuselage) sloping at the same angle. In contrast, the YF-22 had twin fins/rudders with thick inboard sections, inclined at just over 60° and operating in conjunction with horizontal tails. The overall arrangement had much in common with the Su-27, but what made it unique is that the wing came back past the vertical tails and even past most of the horizontal tailplanes; in other words, the latter were recessed into the trailing edge of the wing. You pays your money and you takes your choice; Lockheed would claim better control, using the horizontal tails to boost response in roll as well as pitch, while Northrop would claim lower weight, drag and radar signature. All the offered engines could have two-dimensional nozzles, giving limited thrust vectoring, and both aircraft had air-refuelling receptacles.

Traditional radars usually have the antenna housed in a transparent radome. If the radome is electronically transparent then enemy radars 'see' whatever is inside, and this is likely to be extremely unstealthy. Modern fighters, such as the rival ATFs, do not have mechanically steered antennas housed inside radomes at all. Instead they have what are called 'active-array radars' (see the author's *Avionics*, another PSL book). Such radars have hundreds of apertures through which emerge the radar signals separated in time by computer-controlled phase-shifters which steer the beam in any given direction. In addition to active-array radar the ATFs introduced an infra-red search and track system (IRST), which is an obvious essential for any modern fighter and had been in service since 1982 on the MiG-29, and even longer on the F-106 and F-14. In August 1990 the USAF said it had 'eliminated' the IRST, on the grounds that 'the technology is not ready for the first ATFs'! Both ATFs had secondary attack capability, in some cases using stores carried externally. Primary air-to-air weapons, the AIM-9 and AIM-120, were carried internally. The Northrop had a very large bay along the centreline, while the Lockheed carried its AIM-9 dogfight missiles in extra bays in the sides of the fuselage, outboard of the inlet ducts.

There followed a major battle, not only between the rival ATFs but also, in a sharply changed world situation, between the ATF programme and Congress. In the first edition of this book I said that I regarded the announced unit price of an ATF, $48 million, as mere wishful thinking. In fact it was to be 15 August 2001 before the production go-ahead was signalled to the successful bidder, Lockheed. The F-22A Raptor was actually an almost complete detail redesign, compared with the YF, though the engine remained the Pratt & Whitney F119, rated at 38,000lb thrust, fitted with nozzles capable of vectoring 20° up or down. It took years for the USAF to admit that this could actually be useful in flight, in enhancing manoeuvrability. Deliveries, which began in August 2003, had by mid-2008 reached 121. To show how wide of the mark the original price estimate was, the Fiscal Year 2007 budget gave the unit weapon-system cost as $171.8 million. Whether funding can also be found for a long-discussed FB-22 bomber version, which

might at last enable the B-52H to be retired, is looking increasingly problematical.

In 1969 Taiwan established the Aero Industry Industrial Corporation (AIDC), and in 1982 this launched a programme to create an indigenous fighter, named Ching-Kuo. Assisted by such partners as General Dynamics, Garrett (later Honeywell) on the 9,400lb afterburning F125 engine, and Westinghouse on the radar, the prototype flew on 28 May 1989. Resembling an F-16, but with the two engines fed by Hornet-style inlets, a total of 102 were built, plus 28 tandem-seat trainers, ending in 1999. Further production was replaced by purchases of F-16s.

A word should be said here about the forward-swept wing (FSW). Such a wing offers certain benefits, to high-subsonic as well as supersonic aircraft. The farsighted German aerodynamicists studied such wings during the Second World War Two, and Junkers even planned an FSW bomber (and flew a crude test vehicle, the Ju 287), but at that time such wings were impractical. A moment's consideration will reveal that, at high speeds, an FSW is unstable; the slightest disturbance will cause it either to rip itself off the aircraft, or to cause a violent uncommanded rotation in pitch. The key requirements to the achievement of an FSW aircraft were the rigidity and light weight of composite (e.g. carbon-fibre) construction, and the instant response of computerised FBW flight controls.

The advantages of an FSW are fundamental. With traditional swept wings, there is always a powerful tendency for the air to flow spanwise, outwards towards the tips. Trying to arrest this with fences causes drag, and all other palliatives introduce additional problems. The result is that conventional ailerons become less effective, while the progressive increase in lift coefficient towards the tips almost always means that, as AOA is increased, the tips stall first. Loss of lift at the tip (the rearmost part of a swept wing) starts a chain reaction which causes violent pitch-up, and various penalties are suffered in designing the aircraft so that this does not happen. Another factor is that the sweep of the leading-edge shock is less than that of the leading edge itself, increasing supersonic wave drag. With the FSW the shock is obviously swept more acutely than the leading edge, giving reduced drag. Flow tends to be inwards, giving perfect aileron effectiveness. The tendency for the root to stall can be countered by various methods, including adding a small aft-swept root section, and providing downwash from a foreplane. Both

So far the two Grumman X-29s are the only supersonic FSW aircraft to have flown. Will this be another idea that does all it was claimed to do and then sinks without trace? *PRM Aviation*

these methods were used in the Grumman X-29A, the pioneer FSW aircraft. The FSW can have higher aspect ratio, and thus be more efficient, and have reduced leading-edge sweep, compared with a traditional swept wing with the same area, taper ratio, shock sweep and bending moment. Alternatively, for the same span, area, and shock sweep, the FSW moves the aerodynamic centre inwards towards the root, significantly reducing bending moment. Altogether, the results are a smaller and lighter wing, enhanced manoeuvrability (with virtually spin-proof qualities), and, especially, improved handling at low speeds, reduced stalling speed, and lower drag across the entire speed range, which means higher speed and climb, or else a smaller engine burning less fuel.

To make the FSW practical it was necessary to make the wing skins from fibre (carbon) reinforced composite sheets assembled in layers with the alignment of the fibres chosen to give the wing the immense stiffness needed in the required direction. The X-29A was designed to be supersonic, and its wing has t/c ratio of 6.2 per cent at the root and 4.9 per cent at the tip. To make such a wing proof against divergence using traditional metal would have required that it be almost solid. Using powerful computers to determine the best answers, the X-29A wing was skinned with plies where the carbon fibres were mostly aligned about 9° forward of the leading-edge angle. When this wing deflects upwards under air loads, as in high-g manoeuvres, the outer parts bend upwards without change in AOA, so there is no tendency to divergence. This wing had a forward sweep at 25 per cent chord of almost 34°. The leading edge was fixed, while the trailing edge was formed by flaperons used for roll control and for high lift at high AOA. The root was extended to the rear in a large strake, which terminated in a powered flap. This flap, together with the sharply tapered powered foreplane, was used for control in pitch. The degree of instability of the X-29A was 35 per cent,

so the pilot was totally dependent upon the triply redundant FBW controls. He could not fly the aircraft unaided for a split second. NASA used the two X-29As for many years at Dryden Flight Research Center, adding a programme requested by the USAF in which small jets blasted from the top of the nose were used to create giant vortices.

Grumman's unsuccessful competitor in the FSW programme was Rockwell, successor to North American and eager to regain some of its lost position as a world leader in fighters. In the 1970s Rockwell studied problems of manoeuvre, using supersonic wing and control designs, and gathered the information in a pilotless demonstrator called HiMAT (for highly manoeuvrable aircraft tech-

The Rockwell/MBB X-31A was designed purely to assist in making future fighters more manoeuvrable. It has three jet-deflection paddles, but no means of making the decoupled manoeuvres demonstrated from 1982 by the F-16 AFTI. *PRM Aviation*

nology). Although it had aft-swept wings, HiMAT was the first flying machine to use composite ply skins tailored so that, as the wings flexed under severe flight loads, the AOA varied in the most favourable manner. In 1984 Rockwell teamed with MBB of Germany to build a successor, this time piloted, designated X-31A. The programme was also known as Enhanced Fighter Manoeuvrability (EFM). It was particularly aimed at solving the problems of post-stall manoeuvring, in which a fighter can continue to engage in controlled combat, and aim a gun, with AOA up to 90° or more, which the MiG-29 and Su-27 are almost achieving today. The X-31A was a small aircraft, powered by a GE F404 engine with afterburner. The configuration was the fashionable one of an aft delta wing and high canard foreplane, the engine being fed by a rectangular ventral inlet with a hinged lower lip. Honeywell provided the flight-control computers, just as it did for the X-29A. The entire leading edge was hinged and powered, and the trailing edge

comprised two flaperons on each side. The foreplane was very close to the nose, and was a single slab on each side. There was a single rudder, an airbrake on each side and, importantly, the afterburner nozzle was surrounded by a secondary nozzle carrying three large powered vectoring flaps with which the jet could be deflected. Gross weight was 15,935lb, and its maximum Mach 1.3. The first of two X-31As was rolled out on 1 March 1990. More than 400 hours were flown exploring the possibilities. Strangely, the X-31A could not make the decoupled manoeuvres that the AFTI F-16 could.

Opposite: Sunlight makes this F-22A Raptor appear gold-coloured. The engines are in full afterburner.

Below: Two F-22 Raptor interceptors, seen from an air-refuelling tanker. *Both courtesy Lockheed-Martin*

Chapter 12

AEROSPACE-PLANES

It seems appropriate in this final chapter to take a look at the fastest aircraft of all. These are not merely supersonic but hypersonic, generally taken to mean capable of Mach 5 or above. Such speeds are associated only with flight levels high above the Earth, where the air density is extremely low. This takes us into a new realm of flight. Here the highest clouds seem remote, far below. Above is a sky bearing an odd resemblance to that on a clear night, such a dark violet-blue it looks almost black. If we go to about FL1500 (150,000ft) there are so few particles to scatter the sunlight that, for all practical purposes, we can say the sky is black.

Despite the awesome speed we do not seem to be moving very fast. Looking down from our Mach 5 vehicle at 150,000ft our rather remote planet appears to go past beneath us at exactly the same rate as it did from Concorde at 60,000ft, from an ordinary jetliner at 25,000ft and from a Tiger Moth at 4,000ft. It also happens to look the same speed as from a spacecraft in an orbit at 150 miles.

Hypersonic flight above 99.99 per cent of the atmosphere is often called superaerodynamics. Despite this different name, there is no difference in principle between hypersonic flight and supersonic flight. All the amazing reversals in behaviour took place at Mach 1, and Mach 15 is just like Mach 1.5, only much more so. On the other hand, there really are some very big contrasts. For example, enthusiasts who are excited by fighters may not be aware of the way total pressure – the dynamic or ram pressure – builds up as speed increases. At Mach 1 the pressure (inside an efficient engine inlet, for example) is less than double that of the surrounding atmosphere. An F/A-18 at full throttle at about Mach 1.8 would generate a pressure of about 5.7 times atmospheric; an F-15 at Mach 2.5 experiences just over 17 times, and a MiG-25

at Mach 3 hits 36.73! At hypersonic speeds the pressures are hundreds of times greater.

If a MiG-25 tried to fly at Mach 3 at sea level, the pressure in its inlet ducts would reach 540lb/sq in, three times the pressure in the original boiler of the *Flying Scotsman* locomotive. I don't need to check with the Mikoyan team to calculate that the MiG-25 would explode under these conditions. It is for this reason that hypersonic aircraft fly very high. Indeed, above 150,000ft the atmospheric pressure is so extremely low that, even when multiplied hundreds of times, the dynamic pressure, or q, experienced by the aircraft is also low. As we shall see later, on the X-15 q could be 1,000lb/sq in on a flight profile selected for speed, but only 1lb/sq in on a high arching trajectory, but with speed still well over 5,500ft/ sec (3,750mph or Mach 5.7).

This leads to a very important fact. In hypersonic flight the dynamic pressure can be so low that ordinary aerodynamic controls are ineffective, and for full control of the trajectory the aircraft – or, rather, aerospacecraft – has to be fitted with some form of reaction-jet controls similar to those fitted to the Harrier. In the latter aircraft such controls are needed because, despite the high air density, the speed can be very low or even zero. In the case of the aerospace-plane it is just the opposite: despite the tremendous airspeed the atmospheric density is close to zero, so the forces generated by traditional controls are insufficient.

On the other hand, thermal considerations are likely to be serious. Kinetic heating does not vary with the airspeed, but with the square of the airspeed, and in hypersonic flight the skin temperatures are likely to reach frightening levels quickly. Although the pilot could not see it, the q-ball on the nose of each X-15 glowed brilliantly. As in the case of the ratio of dynamic to static pressure, there is nothing any designer can do about kinetic heating, but again the problem is alleviated by flying very high. When the surrounding atmosphere is almost a vacuum, the rate of heat input is quite low, so in sustained flight it may be possible to stabilise temperatures by dissipating heat by black-body radiation. If the time spent at maximum speed is not very long – say ten minutes – an alternative answer may be to use heat-sink methods of construction, with solid metal leading edges, and unusually thick skins, able to absorb most of the heat energy, or to apply an ablative coating before each flight. Both techniques were used in the X-15 programme.

If there is a fundamental difference between ordinary aeroplane flight and aerospace-plane flight, it is due to the very low density of the atmosphere at the cruising altitude. This exerts a large influence on the mean free path, which is the average distance travelled by any gas molecule – in air, probably a molecule of nitrogen or of oxygen – between successive collisions with other molecules. This distance is millions of times greater than the average distance between the molecules, but at sea level it is still a very short distance, in the region of three millionths of an inch, or 1/100,000 of a centimetre. But as we climb up to and through the stratosphere the mean free path gets longer and longer. At a height of 50 miles it is in the order of one-eighth of an inch; at 60 miles over an inch; at 70 miles 1 foot; and at 90 miles 10 feet. This change does have practical repercussions on superaerodynamics.

The main factors in determining the flow around a high-speed object are the object's own size and shape, the speed of the body, the speed of sound, the ratio of the specific heats for air, and possibly the absolute temperature. None of these factors can vary very much, but the mean free path can vary by a factor of many millions, and this does change the conditions. As we fly our Tiger Moth or 747 we have R (Reynolds number) up in the millions, and M (Mach) less than one, so that M is insignificant in comparison with R. Reynolds number is an expression of the size of a body in relation to the flow past it, numerically equal to air density multiplied by velocity multiplied by a characteristic length (usually the chord of the wing), divided by air viscosity. It is the most fundamental characteristic describing flow round a body. Our aerospace-plane experiences totally different conditions in which M is much greater and R much smaller. Instead of one being millions of times greater than the other, the two are comparable; indeed, M may even exceed R.

This alters what happens. The first thing to note is that our ideas about boundary layers go out of the window. As long ago as 1879 James Clerk Maxwell showed that when M and R become comparable, in other words when the mean free path is similar to the thickness of the boundary layer, the notion of a boundary layer breaks down. The molecules in contact with the body are no longer at rest with respect to the surface, but slide across it. Such a regime is called slip flow. Secondly, when air density is exceedingly low, and M similar in magnitude to R, we can no longer ignore the thickness of the shockwave. Such waves are several mean free paths thick, because each molecule of gas requires several collisions to reach the new values of pressure or temperature appropriate to the region downstream. Each molecule has five degrees of freedom: three translational (up/down, left/right, to/fro), one vibrational and one rotational. As mean free path becomes significantly long, we have to notice that, for every molecule, each degree of freedom has its own temperature and its own shockwave thickness, which makes everything more difficult to calculate and almost impossible to portray in a sketch.

All this just shows that we cannot take too much for granted, and that in many respects aerospace-planes are a race apart. They form a bridge between the familiar aeroplane and the equally familiar tail-standing one-shot space rocket. What makes their design harder is that they have to fly like both, glowing with the friction at several thousand miles per hour in the black emptiness of near-space and a few minutes later lowering landing gear to touch down on a runway and taxi to a terminal. We have not yet achieved the last bit.

The first man-made object to go out at hypersonic speed into near-space was the German A.4 (so-called V-2) rocket of the Second World War. This stood vertically on its tail on a launcher that could be rotated. When the target had been selected, the launcher was carefully aligned with the exact launch azimuth (bearing). The rocket initially climbed vertically, but soon was commanded, by aerodynamic tail fins and by carbon vanes in the jet, to tilt over towards the target until it was inclined at about 40°. Then, at a precomputed time calculated to give the exact cutoff velocity, the rocket was shut off. From that point, the A.4 was a dumb ballistic device, no better than a cannon ball. Its kinetic energy kept it arching up to apogee (peak altitude) about 60 miles above the Earth. Then it simply began to fall again, its speed at this point having fallen off, though it was still

A German A4 rocket (better known as the V2) ready for firing at Peenemünde in 1944. This was by many years the first hypersonic winged vehicle. *Jonathan Falconer*

would have been likely to be detonated by the heat. The author always wondered why they put the warhead in the nose, which on the descent was the hottest place. In the lower atmosphere the tail fins would have been highly effective, but no guidance was forthcoming, because no suitable guidance method existed at the time.

In January 1945 the Germans fired two A.4b rockets, fitted with a pair of large sweptback wings. The first failed, but the second climbed above the atmosphere at 2,700mph (Mach 4.1), to become the first winged vehicle to fly faster than sound, though a wing failed on re-entry to the atmosphere. Ostensibly the A.4b was to stretch the rocket's range, so that London could still be hit despite the Allied occupation of the original launch sites. In fact Wernher von Braun and his team had secretly longed to build spaceships, or rather aerospace-planes. The project drawings at Peenemünde (hastily put away if any visitors arrived) showed the warhead replaced by a cockpit, and a tricycle undercarriage. He wrote in (*Project Satellite*, Wingate, 1958), 'We computed that the A.4b was capable of carrying a pilot a distance of 400 miles in 17 minutes. It might have taken off vertically like an A.4, and then landed like a glider on a medium-size airstrip.'

As far back as 1936 the test stands at Peenemünde were designed to handle rockets of 200 tonnes thrust, and before the war the basic design had been prepared of the A.9/A.10 combination. A.10 was to be a huge rocket with an engine burning nitric acid and diesel oil to give a thrust of 440,900lb. Slotted into the top of the A.10 was to be the A.9, rather smaller than an A.4 and fitted with full-length wings, set at 90° to a pair of tail fins. This would separate from the A.10 at a height of about 37 miles, and then climb under its own power, reaching a cutoff velocity of 6,200mph (Mach 9.4). After reaching apogee, the A.9 was then to plunge back into the atmosphere but, unlike the A.4, it retained guidance. Its control fins were to pull it up and back into a zoom climb. Thus, by means of repeated 'skips' into and out of the atmosphere, it was calculated to fly over 3,100 miles (i.e. it could cross the North Atlantic). The

supersonic. Because of the extremely low air density, the rocket had almost no weathercock stability, despite its huge tail fins, so for a minute or two it would be plunging down travelling more or less sideways, though still exactly aligned with the target. As the atmosphere grew denser, the fins would tilt it point-downwards. The Germans had no way of knowing how each rocket was behaving, but common sense suggests that what happened was that the nose-down rotation would continue until the rocket was flying almost sideways pointing back towards the launching point. Then it would swing round again, each swing being much less in magnitude as the air rapidly grew denser, the speed building all the time. By the time the rocket was plunging back through the stratosphere its tail fins would have forced it to fly point-first. At 3,500mph (Mach 5.3) the nose would glow visibly red, but thick glass-wool lagging insulated the interior. The warhead was filled with nearly a ton of Amatol. Much more powerful explosives could have been used, but these skips were at the expense of the vehicle's kinetic energy, each skip being slower than the previous one. During the Second World War the Peenemünde team had to work on A.10/A.9 as a weapon to be launched from Brittany to put warheads on New York or Washington, but what von Braun dreamed of was his A.10 making a normal landing on an American runway.

Sir Frank Whittle told the author, 'Inventing the turbojet was nothing. The important task was making it work.' In contrast, the German rocket pioneers were gifted with wonderful strategic vision, but were light-years away from making their dreams reality. Even today, 50 years later, we still have nothing like A.9/A.10, and yet this was not the greatest of the German pioneer aerospace-planes. Back in 1938 a considerably superior concept was worked on in Vienna by Eugen Sänger and his future wife Irene Bredt. Sänger was an engineer and Bredt a mathematician, so together they had the capability not only of having a bold idea, but also of carrying out the outline design and all the necessary calculations. After the war began, it had to be recast as a weapon, and its popular name was the Sänger-Bredt antipodal bomber, because its calculated maximum range of 14,596 miles would have enabled it to bomb any place on Earth.

The Sänger-Bredt vehicle does not look too odd today, except that a twin-finned tail is out of fashion,

but in 1938 the concept was staggering. Its length was to be 91ft 10in, the span 49ft 3in, and its starting weight 100 tonnes. It was to take off at a frightening Mach 1.5 from a huge captive rocket-propelled truck riding on a 1.8-mile-long monorail. After climbing at 30°, the low-wing vehicle was to reach orbital height of 100 miles at a speed of 6km/sec (13,422mph). Bredt calculated that it would then re-enter at such a shallow angle that it would skip out into space again, like a flat stone skipping across a pond. She calculated that it would have covered 7,643 miles after the fifth skip and 9,818 after the ninth. Somewhere en route it would drop a 661lb bomb. Finally, when it was down to 24.8 miles altitude, the vehicle would enter the final glide to make a landing on 'the other side of the world' at 90mph With only slightly more propulsive energy it could make a complete orbit, landing back at the launch site.

The antipodal bomber was abandoned in 1942, because even Nazi Germany occasionally recognised when something was incapable of achievement in the current state of technology. At the same time, this project was completely sound in conception, and several German companies saw fit to display large models of it at Paris and Farnborough airshows. It led to several postwar projects, such as a detailed proposal for a man-carrying A.4 by R.A. Smith and H.E. Ross of the British Interplanetary Society in 1946, and a boost/glide hypersonic rocket

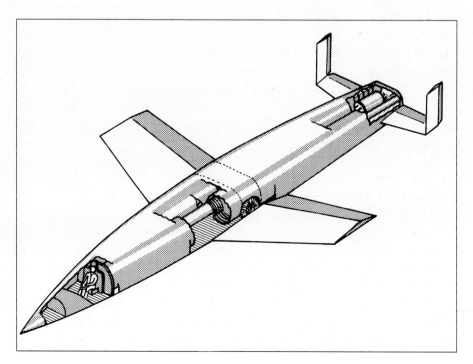

An original wartime sketch of the Sänger-Bredt 'antipodal bomber'. Clearly there were two tubular propellant tanks, but it is not clear how the 'rocket motor and auxiliary machines' functioned. *via Ken Gatland*

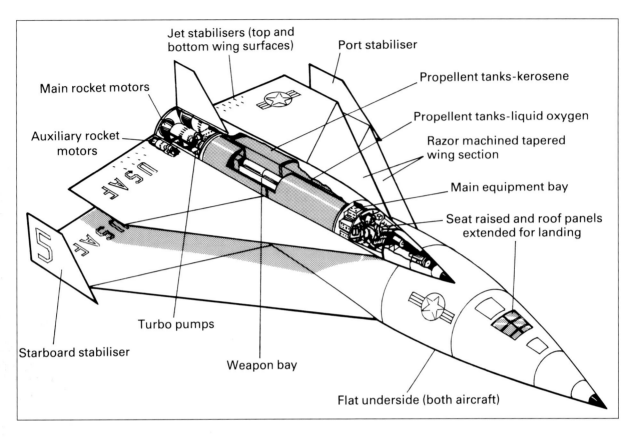

Jet stabilisers (top and bottom wing surfaces)

Port stabiliser

Propellent tanks-kerosene

Main rocket motors

Propellent tanks-liquid oxygen

Auxiliary rocket motors

Razor machined tapered wing section

Main equipment bay

Seat raised and roof panels extended for landing

Turbo pumps

Starboard stabiliser

Weapon bay

Flat underside (both aircraft)

The Dornberger/Ehricke BoMi (bomber missile) worked on at Bell Aircraft would look quite futuristic today. British artist John W. Wood did this drawing in 1954!

glider by Dr C. Hsueh-Sen of Caltech in 1949. Especially interesting was a series of two-stage vehicles proposed by Dr Walter Dornberger and Krafft Ehricke. They had been top men at Peenemünde, Dornberger, as a major-general, being Commandant. Their secret wartime ideas for winged passenger carriers or bombers culminated in more detailed design concepts in the 1950s, the most important being the BoMi (Bomber Missile), which was worked on at Bell Aircraft in 1951–5. This had the form of a tailless delta aerospace-plane riding on the back of a much larger tailless booster with twin fins mounted on a wing tapered on the leading edge. All parts were to be reusable, the booster and upper component blasting off vertically, and separating at supersonic speed at about 20 miles. Both boost/glide and orbital upper stages were studied, including a passenger-carrying version.

The manager for the BoMi version was the USAF Air Research and Development Command, later called Systems Command.

With hindsight, of course something like BoMi could have been built and flown, but neither the technology nor the will to find the money existed at that time. Something more immediately practical was being worked on at Douglas Aircraft's El Segundo plant, where the Chief Engineer was Ed Heinemann. Perhaps the greatest designer of combat aircraft in history, Heinemann had already created the D-558-I Skystreak and D-558-II Skyrocket (Chapter 6). Before the end of 1953 the latter had exceeded Mach 2 and reached 83,235ft. This might have made some designers feel quite pleased, and content to rest on their laurels. In contrast, Heinemann realised that rocket propulsion had already taken ordinary aeroplanes to twice the speed of sound at the edge of airless space. With a little more input energy, such as higher thrust and bigger propellant tanks, the flight envelope could be opened out to a fantastic degree. With virtually no air to cause drag, the speed could climb right up the Mach scale, and the kinetic energy could take the vehicle to heights never before dreamed

Ed Heinemann always regretted that the Douglas D-558-III was never built. It promised to outfly the X-15 five years earlier.

of. Heinemann said 'Let's go for a million feet.' So work began on the first immediately practical aerospace-plane, the D-558-III.

The US Navy Bureau of Aeronautics funded the resulting report, dated 28 May 1954, which described the Dash-III, or Douglas Model 671. It was not a large aircraft – or rather, aerospace-plane – with a span of 18ft, a length of 47ft and a weight of 22,000lb including 15,000lb of rocket propellants. The wing, of 4.5 per cent thickness, and the rather thinner tail, were not swept but tapered on the leading edge. For the first time a designer elected to use a blunt, squared-off trailing edge on the wing and tail. The idea of a single-wedge aerofoil (see diagram in Chapter 3), with a sharp leading edge and squared-off trailing edge, at first glance seems ridiculous because it is so unlike anything we are used to, even on

supersonic aircraft. Yet it was explored mathematically by Ackeret and others more than a century ago, and is entirely sensible at high Mach numbers. Even if the suction on the base (trailing edge) were a vacuum, it would still pay to use a single-wedge profile for t/c ratios greater than 8 per cent at Mach 4. In fact, the base pressure is typically about 35 per cent of static pressure, so at Mach 4 the single-wedge is superior to ordinary sharp trailing edges at t/c greater than 5 per cent or even slightly less. On the Model 671 Kermit van Every elected to use a biconvex wing rather like that of the Skyrocket, but with the trailing edges squared off to give the minimum combination of weight and drag, with acceptable landing characteristics.

The engine was to be the Reaction Motors XLR30, rated at 55,000lb at sea level and at about 60,000lb in near-

X-15 slung beneath its parent aircraft, the venerable NB-52A serial 52-0003. *PRM Aviation*

vacuum conditions. Fuselage tanks held the propellants: liquid oxygen and anhydrous ammonia. There were plenty of problems. Skin temperatures were calculated to reach 1,400°F (760°C), and the answer finally adopted was to make most of the skin in magnesium 0.75 in thick, serving as a heat sink. The leading edges were to be of solid machined copper. Ed said 'In the dark of space, the leading edges will glow quite brightly; fortunately, the

pilot won't be able to see them.' The D-558-III was to be dropped from a B-52 at Mach 0.75 at 40,000ft. Once well clear, the pilot would fire the engine, and pull up into a climb at 38°. The XLR30 was to burn for 75 seconds, after which the Dash-III would continue in a ballistic trajectory to an estimated 700,000ft. Small hydrogen peroxide rockets in the tail and wingtips would provide jet-reaction control to keep the vehicle pointing the way it was going. Having arched over at apogee, it would then plunge down again, re-entering the atmosphere and reaching a peak speed of about Mach 9 (almost 6,000mph). An advanced autostabilising autopilot and powered controls would be needed. The cockpit would be thermally insulated by a two-inch blanket, while the two small windows were to be double layers of half-inch quartz, with a quarter-inch gap. A special escape capsule would be needed, together with reliable positioning for the final landing back at Muroc (later called Edwards) Dry Lake.

Sadly, senior Navy personnel thought Mach 9 was not their business, and Douglas never succeeded in getting funding to turn the Dash-III into hardware. But it created great excitement with the Air Force, and with the NACA, which by 1954 was already trying to put together a programme for an aerospace-plane. The original spur for this proposal was the Bell/Dornberger study already mentioned, but the NACA was not keen on tail-standing vertical launch. During 1954 the NACA at Langley began a research programme specifically aimed at solving the basic problems of an aerospace-plane. While this was gathering momentum, on 5 October 1954, a meeting was held with the USAF and Navy at which it was agreed to solicit bids for a vehicle to fly at 250,000ft

at 6,600ft/sec (4,500mph or Mach 6.8). Nine companies responded, and the finalists were Douglas and North American Aviation. Indeed, the Navy BuAer placed Douglas significantly in front, but after much further discussion – which even involved stretching out the timescale by eight months after NAA had said it wished to withdraw, because of pressure of other programmes – the job was given to NAA on 30 September 1955. Even then the objectives were well short of those Douglas had aimed at with the D-558-III. Heinemann was peeved.

Previously known as Project 1226, the NAA aircraft was assigned the designation X-15. NAA received a contract to build three, serial numbers 56-6670 to -6672. Aerodynamically the design was straightforward, with

The third X-15, 56-6672, just after ignition of its XLR11 rocket engines. The photograph was taken from the NB-52A. *PRM Aviation*

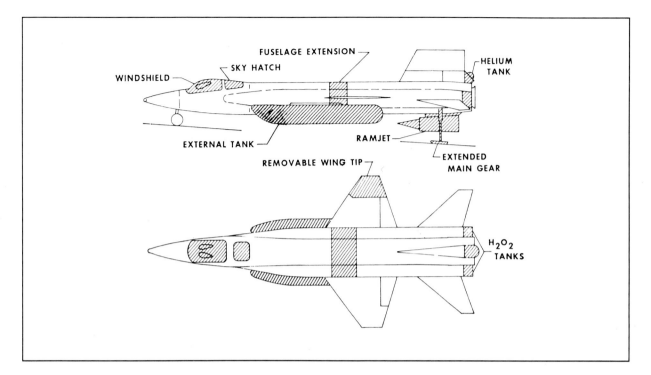

Major modifications resulted in the X-15A-2, but the heavy A-2 never reached its design performance of Mach 8.

a long cylindrical fuselage occupied mainly by tankage, stubby tapered wings, and a fairly conventional tail, but there were many interesting features. Whereas the wing and horizontal tails had a conventional almost biconvex 5 per cent section, with a modest blunt trailing edge (on the wing this squared-off edge tapered from 2.125in deep at the root to 0.375in at the tip), the vertical tails were unique. First, they comprised almost rectangular surfaces, tapered on the leading edge, above and below the rear fuselage, the main part of the ventral fin being jettisoned before landing because it would have fouled the ground. Second, the main part of the upper fin was pivoted as a single unit, serving as a slab rudder. Third, the lower part of the upper fin opened out at 45° on each side to form the speed brakes. Fourth, and most remarkable of all, the aerofoil profile adopted was a single wedge, of 10° leading-edge angle, the trailing edge thus being more than 12in thick and completely square to the airflow! We have already seen how at high Mach numbers the blunt single wedge is a preferable section, and NAA designer

C. H. McLellan computed that the seemingly huge high-drag shape would actually have less drag at hypersonic speeds and give much better directional stability and control.

Yet another odd feature was that, while the wings were unswept and tapered, the horizontal tails were swept (45° at 25 per cent chord) and tapered. They were arranged to operate as tailerons, providing the sole control in roll. Not least, they were mounted only an inch or two lower than the wings. Previously aerodynamicists had 'known' that you never put a tailplane in the wake of a supersonic wing, far less a hypersonic one, because there are violent variations of total pressure, Mach number and everything else, as you traverse the few inches from above the wake to below it. The author was amazed when he saw the X-15 horizontal tail location, notwithstanding the fairly sharp anhedral. A further complicating factor, though at subsonic speed, is that just in front of the tailerons the flaps came down to 40° before landing. During the early planning the design team, led by Harrison Storms, who later masterminded the Apollo lunar vehicle, considered the same engine as planned for the D-558-III. Eventually they decided on the totally new Reaction Motors XLR99. As it was clear the airframes would be ready long before this engine, it was agreed to complete them temporarily with two superimposed XLR11 engines developed from the primitive type fitted to X-1 and D-558-II aircraft.

Thus, the X-15s were originally built with a forward tank for 861 gallons of lox and an aft tank for 1,203 gallons of alcohol/water mix. The tanks were toroidal, with tubular and spherical tanks down the centre for gaseous helium, though the final feed to the engines was by turbopump. Each engine had four thrust chambers rated at sea level at 2,000lb, so that total thrust at high altitude was nearly 19,000lb, the pilot being able to select from one to eight chambers. Later, with the LR99 engine, the propellants were lox and ammonia.

The small (200sq ft) wings were not attached direct to the fuselage but to large fairings forming chines running almost the full length of the aircraft. These served to carry a mass of pipes, cables, flight-control rods and other items past the tankage. Moreover, they were also planned to contribute to the aerodynamic lift, and at Mach 5 these chines provided about 59 per cent of the lift. Another interesting design choice concerned the landing gear. Douglas had believed it could use wheels throughout,

but the other bidders for Project 1226 accepted that, for reasons of temperature, the main gears would have to be some form of retractable skid(s). NAA began with main gears in the expected location, but in the final design they were right at the tail. Each unit comprised a rigid strut carrying a heavy steel ski, the only shock-absorption being by limited outboard deflection of the struts on impact (like a Cessna). In the nose was the twin-wheel nose gear. All units were locked up before the take-off of the NB-52 parent aircraft, and were released close to touchdown to fall under gravity and lock under air loads. The obvious problem was the enormous sagging load,

X-15 No. 2 after being rebuilt to A-2 standard. Later it was repeatedly given a thick coating of white ablative material, a fresh coat for each flight.

especially in a hard impact of the nose gear, and on one occasion the fuselage buckled aft of the cockpit.

Early in the programme it seemed clear that there was only one possible material for the skin: Inconel X. Closely related to British Nimonic alloys, this high-nickel alloy had previously been needed only in the hottest parts of gas turbines. A vast amount of research was needed to ascertain the best way of using it, and of joining it to an underlying structure which was generally of titanium alloy or, in cooler places, 2024-T4 aluminium alloy. The best titanium alloy, also picked for the Boeing SST, was 6 per cent Al, 4 per cent V (vanadium). Almost all the joints were welded, usually with fusion methods (gas or arc) in an inert atmosphere. The entire completed airframe then had to be cooked in a furnace to relieve stresses and provide essential heat treatment! This was before adding the two windshield transparencies, which at first were quite large rectangles. Made of multiple layers of thick glasses, they naturally suffered severe thermal stress when the outside was heated to 427°C (800°F) while the inside was still cool. Two fractured, the materials were changed, and the glass blocks free-floated in a more flexible titanium frame.

The cockpit was naturally very small, and full of traditional dial instruments which did not include an airspeed indicator. On the tip of the nose was originally an instrument boom, but when the X-15s got down to work this was replaced by a q-ball, a sphere devised by Northrop Nortronics which often glowed bright red with heat. Driven by electronic servos, it was a hydraulically actuated null-seeking device, which indicated the exact direction from which the air (generally highly rarefied) was coming, and indicated q (dynamic pressure). Maj 'Pete' Knight said, 'We have this pseudo-q indicator, and we have inertial velocity and inertial vertical velocity. Airspeed has no meaning to us until we get back into the pattern, and then we use pressure altitude and airspeed subsonically.'

A conventional ejection seat was used, with no front protection against high q, and few pilots really believed it was good for the advertised Mach 4. I spoke to several X-15 pilots, who all said that they would stay on board as long as possible. Another odd feature was that there were three control columns. The main one was used in low-speed atmospheric flight, giving manual control. On the right was a pioneer sidestick, tested on an F-107A, giving wrist control of the same tailerons but via the hydraulic power units. This was soon used almost exclusively, the

central stick being removed. On the left was another sidestick controller, used in hypersonic flight, where the air was too thin (q often less than 1lb/sq in) for the tailerons to be effective. It controlled pitch, roll and yaw by means of 12 hydrogen-peroxide reaction jets, four pairs spaced at 90° around the nose and two pairs at the wingtips. Honeywell later developed an adaptive control system which continuously sensed atmospheric static pressure and dynamic (q) pressure, and automatically selected either aerodynamic or reaction-jet control. Fine in theory, this could become confused and once contributed to a fatal in-flight break-up.

All three X-15s were finished in heat-emissive black, with USAF titles in white later supplemented by NASA in black on a gold stripe across the rudder. The No. 1 aircraft, with two XLR11 engines, made its first gliding flight on 8 June 1959, NAA pilot Scott Crossfield finding the manual controls (at least in the pitch axis) too sensitive. The second flight was made by the No. 2 aircraft, and Mach 2.11 was reached under power on 17 September 1959. Flight 26, by No. 2 aircraft on 15 November 1960, was the first with the XLR99 engine, Mach 2.97 being reached at half-power with the airbrakes open. Altogether the three aircraft made 199 flights, the last on 24 October 1968. The peak altitude reached was 354,200ft by Joe Walker on Flight 91, on 22 August 1963, with No. 3 aircraft. The peak speed was 4,520mph, Mach 6.70, by 'Pete' Knight on Flight 188 on 3 October 1967, with No. 2 aircraft.

The record speed, was achieved after the No. 2 aircraft had been rebuilt following an accident, emerging as the X-15A-2 (see diagram). Apart from the 29in extra section spliced into the fuselage between the lox and ammonia tanks, to make room for a liquid-hydrogen tank to feed a test ramjet (which was to replace the jettisonable lower fin), the main change in the A-2 was the addition of two large jettisonable tanks under the wings. Each some 24ft long, these housed lox on the left and ammonia on the right, extending full-throttle burn time from 90 to 150 sec. Their presence increased gross weight from 34,000lb to 53,000lb, lowering the B-52 drop altitude by over 1,500ft and making the X-15A-2 very heavy and sluggish. Moreover, the pilot had to hold 22° AOA and climb hard as propellants were burned off, to avoid exceeding the 1,000lb q limit on the tanks. Their presence moved the CG down by 9in, and also moved it over 2in to the left of the centreline because the lox tank weighed 8,920lb and the NH_3 tank only

A Boeing artist's impression of X-20 Dyna-Soar about to land. *via Ken Gatland*

6,850. Other changes, apart from the addition of the test ramjet, which was seen as propulsion of future hypersonic aircraft, included extended and strengthened main legs to handle the extra weight and the presence of the ramjet during the landing. The windshield was redesigned with three layers of different material in a small oval frame on each side. The 'sky hatch' was a compartment for a camera or other sensor, whose doors could be opened above the atmosphere. The right wingtip was designed to be replaced either by tips made of test materials or structures, or containing sensors.

Despite the big extension in burn time, the peak Mach expected to be reached did not rise very much. Mach 6 was exceeded by the original aircraft back in 1963, and the highest figure with the help of tanks was only 6.7. The A-2 design speed had been Mach 8, and almost all the shortfall was due to the massive increase in weight

beyond estimates. In fact, it was hoped that the external tanks would result in Mach 7.4 being reached, and revised data showed that the structural limiting temperature of 704°C (1,300°F) would be reached at Mach 6. The answer appeared to be to cover the exterior with an ablative coating, such as had already been developed for ICBM re-entry vehicles. The choice fell on Martin Marietta's MA-25S, a thick white 'paint' consisting of a glass-bead powder suspended in resin. The applied thickness varied up to 0.6in over the wings, canopy and leading edges of the tails, showing that aerospace-planes in this category

FASTER THAN SOUND

are projectiles, whose shape (e.g. the precise aerofoil profile) are often secondary. There were many problems. MA-25S could explode, if soaked with lox and then struck. Equally serious, as it burned – or rather vaporised – it deposited opaque residues on the windscreens. Accordingly, the left screen was covered with a hinged eyelid. As soon as the right screen became opaque, the pilot had no external vision whatsoever until, as speed bled off below Mach 3, the pilot could open the eyelid. This acted as a small canard, causing pitch-up, right roll and right yaw, but at least the pilot then had a tiny area of external vision with his left eye.

You will have gained the impression that X-15 pilots earned their keep. As they accomplished so much, incidentally spending 5hr 59min above Mach 4, and nearly 1hr 26min above Mach 5, I will list them, in the order in which they entered the programme: Scott Crossfield, NAA; Joe Walker, NASA; Bob White, USAF; Forrest Peterson, USN; John McKay, NASA; Bob Rushworth, USAF; Neil Armstrong, NASA; Joe Engle, USAF; Milt Thompson, NASA; William 'Pete' Knight, USAF; Bill Dana, NASA; and Mike Adams (killed), USAF. The two surviving aircraft are on display, No. 1 at the National Air and Space Museum and No. 2 at the USAF Museum.

Although the X-15 never attained its design goals, it was, and remains to this day, in a class of its own as far as aeroplanes are concerned. And, though it was dropped from a parent aircraft and could climb out of the atmosphere, it can certainly be called an aeroplane, unlike the Shuttle orbiter. The plan to test at least one type of hypersonic scramjet (supersonic-combustion ramjet) was never fulfilled, and today there are many other tasks crying out for a manned hypersonic aerospace-plane. Twice it appeared that an American aerospace-plane was quite near to becoming reality. The first was the Boeing/USAF X-20 Dyna-Soar, cancelled by SecDef McNamara who thought it a waste of money. The second was the NASA X-30A National Aero Space Plane, which was abandoned simply because NASA could not afford it.

Earlier in this chapter we left Bell Aircraft, soon to become Bell Aerosystems, studying the BoMi concept. In addition to the piggyback configuration, this was also studied as a tandem vehicle launched on top of various rocket boosters. The key factor making it all possible was the development of rocket engines of unprecedented power, for US ballistic missiles. The first such engine was

designed in 1953 by North American Aviation, which two years later formed a separate division, Rocketdyne, for this rapidly growing work. The first engine was rated at 138,300lb at sea level, burning lox and RP-1 kerosene-type fuel, and was mounted in a triple package as the booster for the Navaho cruise missile (Chapter 6). By 1954 engines rated at 160,000lb were under development for Atlas, while rival Aerojet-General worked on similar engines for Titan. Both of these intercontinental ballistic missiles were considered as possible boosters for BoMi, and other boost/glide vehicles. Such vehicles required several new technologies, including transparent windows able to withstand the searing heat and dynamic q-pressures of hypersonic re-entry to the atmosphere, and a combination of reaction-jet and aerodynamic control systems for atmospheric re-entry and conventional landing on a runway.

While the NACA (NASA from 1 October 1958) was intensely interested in hypersonic boost/glide vehicles from a research viewpoint, the USAF saw immediate military objectives in the fields of reconnaissance and bombing. In 1956 it launched three such projects: Hywards for research, Brass Bell for reconnaissance, and RoBo for bombing. At first largely unrelated, these were integrated in 1957 into three steps of a major programme called System 464L, the vehicle portion becoming X-20 Dyna-Soar (dynamic soaring). Later the system number was changed to 620A. From the start, the project was to some degree vulnerable, and it never enjoyed proper funding, even after the specification and configuration had been agreed, and the prime contract awarded to Boeing on 9 November 1959.

From the start, Dyna-Soar had the form of a small delta, with a bluff, rounded leading edge swept at 74°. The tips curved up into large fins, which in the final design were toed in at the same angle as the leading edge. The wing maintained its full depth back to the blunt squared-off trailing edge, which was formed by two large powered elevons of box form. The fuselage, not blended into the wing, was of approximately square section. The pilot sat on a conventional rocket-boosted seat, in a cockpit with five flat glass windows which at high speeds were protected by a refractory fairing. The aft end of the body was squared off to mate with the booster, there being no on-board propulsion.

Most of the structure was to be of René 41, an alloy of nickel containing large amounts of cobalt, columbium and molybdenum. It was calculated that, during re-entry,

some parts would heat up to 2,371°C (4,300°F). The nose ball and leading edges were studied in reinforced graphite, but eventually were to be of ceramic. The slightly cooler, but still 'white-hot', underside was to be skinned in molybdenum, insulated from the interior by a layer of quartz wool. Six pairs of hydrogen-peroxide jets, above and below the nose, on each side of the stern and halfway along the leading edges, controlled the vehicle above the atmosphere. This reaction system was controlled by the left sidestick, that on the right

First take off by the first wingless lifting body vehicle, the wooden M2-F1 glider created at NASA Ames in 1963.

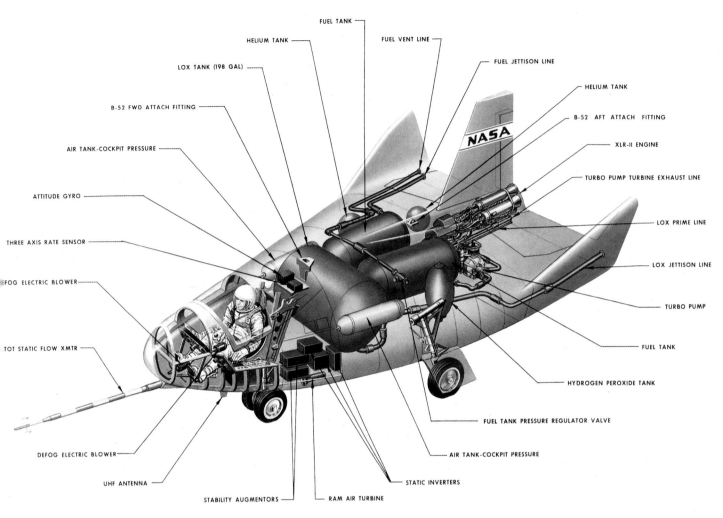

FUEL TANK

HELIUM TANK

LOX TANK (198 GAL)

B-52 FWD ATTACH FITTING

AIR TANK-COCKPIT PRESSURE

ATTITUDE GYRO

THREE AXIS RATE SENSOR

FOG ELECTRIC BLOWER

TOT STATIC FLOW XMTR

DEFOG ELECTRIC BLOWER

UHF ANTENNA

STABILITY AUGMENTORS

RAM AIR TURBINE

STATIC INVERTERS

AIR TANK-COCKPIT PRESSURE

FUEL TANK PRESSURE REGULATOR VALVE

HYDROGEN PEROXIDE TANK

FUEL TANK

FUEL VENT LINE

FUEL JETTISON LINE

HELIUM TANK

B-52 AFT ATTACH FITTING

XLR-II ENGINE

TURBO PUMP TURBINE EXHAUST LINE

LOX PRIME LINE

LOX JETTISON LINE

TURBO PUMP

NASA

With the HL-10, first flown in December 1966, a different shape was explored with a flatter underside and rounded top.

driving the aerodynamic controls; there was no central control column. Immediately before landing, the pilot was to release three free-fall legs, each carrying a ski with an undersurface in the form of a brush of springy steel wire. Avionics were to include inertial guidance and a stability-augmentation system. Span and length were to be 20ft 5in and 35ft 4in respectively, and gross weight 11,390lb. Using a Titan IIIC booster, the velocity at separation was calculated to be 24,490ft/sec (16,700mph, Mach 25.3), and the Air Force changed the requirement from a single to a multi-orbit mission. This demanded the addition of deorbit power. It was proposed to leave the final transition rocket stage attached, so that it could be fired, with the vehicle travelling backwards, in the required retro-fire mode to bring the Dyna-Soar out of orbit. The final stage would then be jettisoned and the Dyna-Soar turned round to re-enter, travelling in the right direction at high AOA.

It was planned to make air drops from September 1964, unmanned launches from May 1965, and manned flights from November 1965. However, in December 1963 Defense Secretary McNamara infuriated the Air Force (again) by cancelling the entire programme. He needed more money for the F-111, and said that similar results could be obtained much more cheaply with unguided capsules, such as Mercury and Gemini.

The latter are outside the scope of this book, because though their attitude was controllable their trajectory was not. Thus, they were 'dumb' ballistic bodies, which on

re-entry fell into the atmosphere like a stone. We have already seen how the kinetic heating experienced by the A.4 rocket influenced the choice of a weak explosive, and the temperatures and q dynamic pressures experienced by a body re-entering from deep space or from orbital velocity are far more severe. When, around 1950, workers began to think about building an ICBM, no way was known of building a re-entry vehicle to protect the warhead. One might have thought that the best answer would be a pointed cone, but in 1951 H. Julian Allen and Alfred J. Eggers (see XB-70 story) discovered the best answer to be a blunt, or bluff, body. It could be faced with an almost flat slab of copper, serving as a heat sink, or be covered with thick ablative material, which would vaporise or char, taking away heat and protecting the underlying structure.

Allen and Eggers were unconcerned about lift/drag ratio, but later in the 1950s workers began to think about how a blunt body might be shaped so that it had

Though powered by almost the same type of rocket engine as the Bell X-1 series the tubby Northrop X-24A was almost 200mph faster. It was controlled by eight powered surfaces – four on the body and four on the outboard fins.

improved lifting qualities. Thus, it might be possible to build a lifting-body aerospace-plane, which could survive re-entry and then be steered – to some degree – so that it could land on a runway. By the time the NACA became NASA, the two workers at Ames Research Center had been joined by others, and come up with a shape called M2. This was a 13° half-cone, the flat side uppermost, with a blunt rounded nose. It achieved an L/D ratio of over 1.4, and when a tapering 'boat tail' was added its

Martin's X-24A was rebuilt in 1971-73 into the long-nosed X-24B, here seen on the NB-52B carrier aircraft. Surprisingly it was only about 100mph faster at 1,164mph. *Philip Jarrett*

stability in subsonic flight became adequate. The shape was refined, with fins added, and in 1963 NASA began testing a simple 1,200lb wooden glider to prove the idea at low speeds. Called the M2-F1 (Manned, modification 2, Fuselage 1), it was designed and built at Edwards, and made over 500 successful flights towed by fast cars or aircraft. This showed that the landing phase was plausible, and by 1963 thermodynamic research in shock tunnels had confirmed that re-entry heat loads would be significantly less than for a winged design. The next step was to build a metal version stressed for supersonic flight. The resulting M2-F2 was built by Northrop, tunnel-tested at Ames in 1965, and flown from 12 July 1966. Length was 22ft 2in, and the twin fins, toed slightly inwards, brought the width to 9ft 7in. Gross weight was 9,400lb. The pilot's feet were in the tip of the nose. Below and behind him was the twin-wheel nose gear, and the main gears retracted inwards on each side.

In parallel, Northrop also made the HL-10, to a shape which had been developed not at Ames but at Langley. Whereas the M shape was round underneath, with a canopy projecting from a flat top, the HL-10 (Horizontal Landing, 10th shape) had a cambered flat bottom, which swept round to form dihedralled fins, a tall central fin and a canopy flush with the rounded top. It made its first glide on 22 December 1966. Northrop also made an adapter, which allowed either type of lifting body to be carried on the X-15 pylon of an NB-52. A typical flight took about 50 minutes to reach the drop point at Mach 0.8 at 45,000ft, and then just four minutes to glide back to land at Rogers Dry Lake. Glide ratio was an unimpressive 3:1, resulting in a descent rate which at first was 250ft/sec at 350mph Sink was arrested to about 10ft/sec at 190mph at touchdown. The HL-10 was fitted with an 8,000lb XLR11 rocket, with which Mach 1.9 at up to 90,000ft was expected. Unfortunately, on the M2-F2's sixteenth flight pilot Bruce Peterson experienced extreme difficulties and crashed. The horrific pile-up formed the opening sequence of each episode of *The Six Million Dollar Man* television series. The vehicle was rebuilt as the M2-F3, with an added central fin. By late 1971 the M2-F3 had reached 89,000ft and Mach 1.7, while the HL-10 had topped 91,000ft and Mach 1.9.

These Northrop-built bodies provided data for a further refined configuration, developed at Ames and contracted to Martin Marietta. The shape was called a 'bulbous wedge', with a flat bottom and three fins. Martin first built three unmanned vehicles in the Prime programme (Precision Recovery Including Manoeuvring Entry). Designated X-24 (company designation SV-5D), these were accelerated by an Atlas ICBM to approximately orbital velocity. They then demonstrated stability, and the ability to manoeuvre under radio command down to Mach 2 at 100,000ft. Martin then built a single body in the Pilot programme (PIloted LOw-speed Test). Designated X-24A (SV-5P), it resembled the Northrop vehicles in being made of aluminium alloy, and had a T-39 nose gear, T-38 main gears with fast pneumatic actuation, and an F-106 type seat. The X-24A had aerodynamic controls only, comprising upper and lower trailing-edge flaps and upper and lower rudders, all with thin blunt trailing edges. The flaps served as elevators, ailerons and airbrakes. The lower rudders provided yaw control, the upper rudders being automatically biased in proportion to q pressure to serve as trimmers. Propulsive power was provided by an

XLR11-13 with four chambers together rated at 8,480lb. Width was 13ft 8 in, length 24ft 5 in, and gross weight 11,595lb.

The X-24A was intended to fill in the portion of flight not covered by the Prime programme, namely from Mach 2 at 100,000ft down to the landing. Following almost two years of ground testing, USAF No. 66-13551 made its first gliding flight from an NB-52 drop at 45,000ft on 17 April 1969. It made 28 flights, the last on 4 June 1971, reaching Mach 1.6 (1,036mph) and 71,407ft. Martin Marietta itself funded two SV-5J lifting bodies intended as pilot trainers. These were similar to the X-24A but powered by a Pratt & Whitney JT12A (J60) turbojet fed from a ventral inlet, but were never flown. One was intended to be modified with a new nose and outer wings to test a new configuration, but in the end it was decided to modify the X-24A. The resulting X-24B made its first gliding flight on 1 August 1973, and the thirty-sixth and last on 26 November 1975. It looked totally different, with a very

One of the first artworks to show what was expected to be the eventual Shuttle configuration. In fact, a totally different concept was adopted, not all of it recoverable

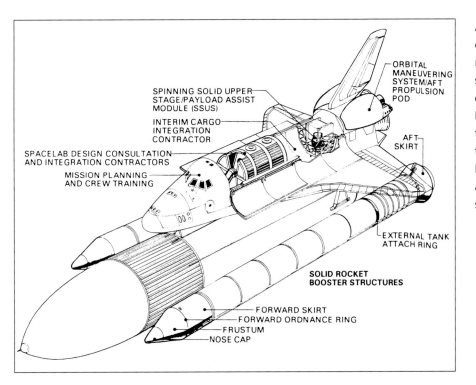

SPINNING SOLID UPPER
STAGE/PAYLOAD ASSIST
MODULE (SSUS)

INTERIM CARGO
INTEGRATION
CONTRACTOR

SPACELAB DESIGN CONSULTATION
AND INTEGRATION CONTRACTORS

MISSION PLANNING
AND CREW TRAINING

ORBITAL
MANEUVERING
SYSTEM/AFT
PROPULSION
POD

AFT
SKIRT

EXTERNAL TANK
ATTACH RING

SOLID ROCKET
BOOSTER STRUCTURES

FORWARD SKIRT
FORWARD ORDNANCE RING
FRUSTUM
NOSE CAP

As the Shuttle prime contractor is always regarded as Rockwell, this shows that quite a lot was contributed by McDonnell Douglas. Grumman made the wing, Convair nearly the whole fuselage, Fairchild Republic the tail and Vought the enormous space radiator panels.

long tapering nose and a plan view of double-delta form, with triangular upward-curved 'wings' added outboard of the fins. The nose gear was replaced by one taken from a Grumman F11F, mounted 6ft further forward. The added wings carried ailerons, interconnected with the upper rudders, the flaps now becoming elevators/airbrakes only. Lifting area was doubled, from 162 to 330sq ft, and the length was increased from 24ft 6in to 37ft 6in. The gross weight was about 13,800lb. The highest speed reached was 1,163mph, Mach 1.76, at about 72,000ft. An X-24C, to reach Mach 8, was never funded.

NASA first sent men to the Moon in July 1969, using the gigantic Saturn V launch vehicle weighing about 6,500,000lb and standing 364ft tall. It was self-evident that to continue using such totally expendable launchers would grossly uneconomic, and later in 1969 President Nixon ordered a look at the next generation of launchers. NASA calculated that switching to re-usable launchers could reduce the cost of putting 1lb into orbit from $1,000 to $20. The result was a plan for a family of vehicles for transport from the Earth's surface to interplanetary space. The first stage, from Earth to near orbit, was to be a re-usable vehicle called a Shuttle. It was to comprise a booster and an orbiter.

Both components could have taken any of several forms. Indeed, one of the first configurations had no

booster, the only parts to be jettisoned comprising huge propellant tanks forming a V-shape enclosing the orbiter. Several arrangements of traditional tandem boost were studied, the orbiter being mounted on top of one or more launch rockets. At least equal interest was taken in re-usable boosters looking very much like the orbiter itself, the latter riding piggyback. Gradually this idea gained ascendancy. From the start, the booster was to be winged, and have air-breathing (probably turbofan) engines to facilitate atmospheric flight and landing. At first it was thought that the orbiter might be a lifting body, but the much greater cross-range distance possible with a winged vehicle was very important. Several illustrations were published showing straight wings, but the final choice was that of a tailless delta, the wing being mounted low on a giant body of almost square section, with a single swept fin on top of an untapered tail end. The prime contract went to Rockwell.

To meet the stipulated payload of 25,000lb, the orbiter came out roughly as big as a Boeing 707, riding on top of a booster about as big as a 747. It was obvious from the start that, the better the system and the cheaper the cost of every subsequent flight, the greater the cost of development. In 1971 a blanket ceiling of $5.15 billion on development showed that the ideal, totally re-usable boost/orbit system was unattainable. Instead, to save

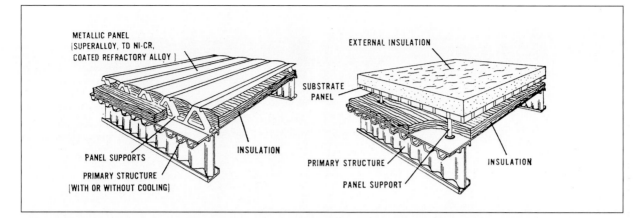

Typical heat-shield concepts for an aerospace-plane.

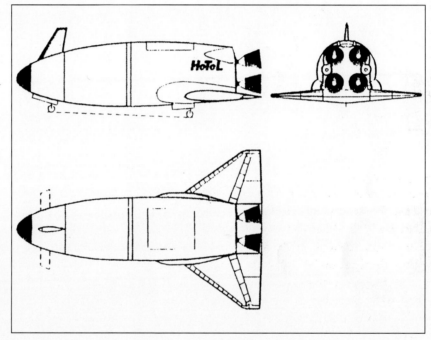

In 1992 British Aerospace published this drawing of a Hotol redesigned for air launch from a Soviet An-225 (with six Trents or eight D-18T engines). The new Hotol is short (121ft or 37m, compared with 203ft or 62m), but weight is up from 196 tonnes to 250. The four engines are Soviet RD-0120 lox-hydrogen rockets of 44,000lb SL thrust each.

development money, parts of the system had to be made expendable on each mission, thus making every launch more expensive. Simple mathematics showed that the extra launch costs would soon swallow up the saving in development cost, but politicians are interested in now, not some time in the future. By late 1971 the final configuration had been reached. The launch was to be vertical, tail-standing like previous rockets. First to reach Pad 39A at NASA Kennedy Space Center would be a gigantic tank, standing over 154ft tall, and with the impressive diameter of 27ft 7in. It would house over 102 tons of liquid hydrogen and 743 tons of liquid oxygen, the usable total exceeding 518,000 gallons. The contract went to Martin Marietta. On each side would be attached

an SRB (solid rocket booster), supplied by Thiokol. Each was to stand over 149ft tall, have a diameter of 146 in and weigh 1,293,246lb (about 586 tons). The thrust of each SRB at lift-off was to be 2,650,000lb, but the rubber propellant, a mixture of ammonium perchlorate, aluminium powder, iron oxide (rust) and a polymer binder, was shaped to reduce thrust by one-third after 55 seconds, to avoid overstressing the orbiter at the time of peak q. Burn time was to be 123 seconds, after which groups of separation rockets (themselves totalling 176,000lb thrust!) would blow the SRBs off the tank, for recovery by parachute into the Atlantic, and subsequent refurbishment. Finally the orbiter would be attached to the side of the tank.

To withstand the white heat of re-entry, the entire underside of the Shuttle Orbiter was covered in hundreds of special refractory tiles. *NASA*

STS-1 *Columbia*, the first Space Shuttle to make an orbital flight, pictured in March 1981, a month before launch on 12 April. *NASA*

As finally built, the orbiter has a span of just over 78ft, a length of just over 122ft and height of 56ft 7in. On Space Transportation System (STS) 26 orbiter OV-103 Discovery weighed 171,419lb and 254,449lb loaded. The one big advantage of the configuration adopted is that the payload is considerably increased, to 29,000lb for a 98° polar orbit and 55,000 to 65,000lb due East.

To anyone used to traditional aircraft design, it seems strange that this, the fastest manned aircraft of all time, should appear to be totally unstreamlined. For example, the wing, of 2,690sq ft, though it has an inboard leading-edge sweep angle of 81°, has a strange bluff profile with a round leading edge which almost immediately reaches the full depth of more than 5ft. Made by Grumman, the wing carries inboard and outboard elevons, which of course are fully powered fly-by-wire. The fuselage, made

basically by Convair, comprises three portions. The bluff nose section contains the crew compartment, with the commander and pilot, a mission specialist, and a payload specialist on the top deck. Three more crew were on the mid-deck. The 60ft central portion is the payload bay, the entire top of which opens in orbit for delivering or recovering spacecraft and other payloads, using a 60ft manipulator arm (0, 1 or 2 arms can be installed). The third section houses the three SSMEs (Space Shuttle main engines). Made by, Rocketdyne, these are high-pressure LH_2/LO_2 engines controllable from 65 to 109 per cent rated power, the thrust at 109 per cent being 417,300lb each at sea level, and 513,250lb in vacuum conditions. Like everything aboard, the SSMEs are designed for repeated use and airline type maintenance,

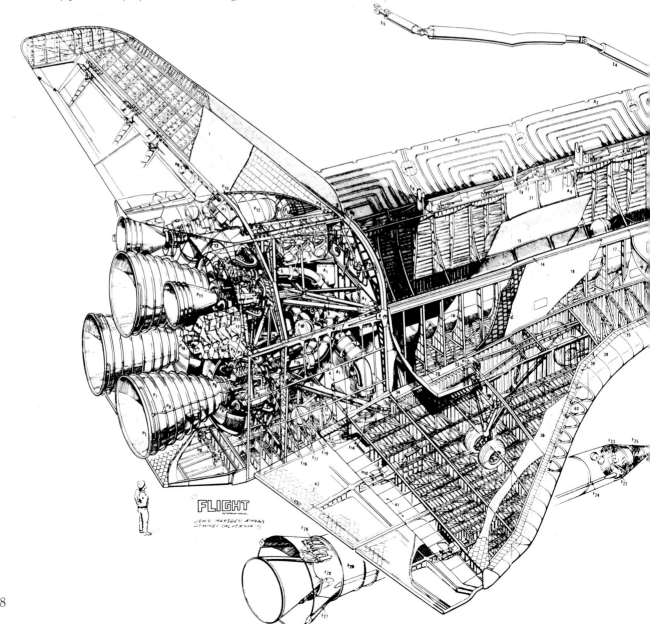

the goal being 7hr 30min operation with 55 separate starts. The orbiter also has two 6,000lb manoeuvring engines, burning nitrogen tetroxide and hydrazine, and 38 Marquardt reaction-control engines grouped round the nose and tail. The vertical tail carries a split rudder which can open to act as a brake, while the underside of the rear fuselage is a large body flap which protects the main engines during re-entry and also serves as a pitch trimmer. The twin-wheel landing gears have main and backup extension systems, but no means of retraction once extended (too bad if they extend in orbit).

At lift-off the SRBs and SSMEs are at full power, generating 7,781,400lb thrust. The SRBs are jettisoned at 120–123 seconds, at a height of 27 miles, followed by the tank at about 8 minutes at 68 miles, shortly before orbit insertion. Missions could last up to 30 days. Retrofire initiates re-entry, the attitude then being altered to re-

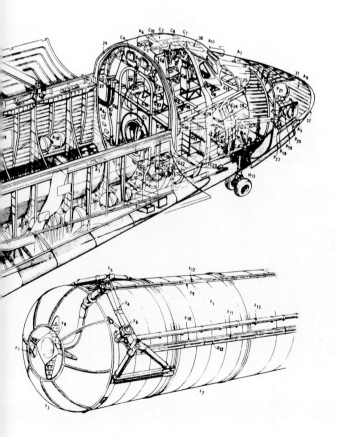

Cutaway drawing of the Shuttle Orbiter.
Flight International

enter wings-level at a high AOA. Here the thermal protection system (TPS) comes into play, to enable the orbiter to withstand temperatures up to almost 1,648°C (3,000°F) on the nose and leading edges. These areas are covered with tiles of reinforced carbon-carbon (RCC) up to 3in thick. The rest of the exterior is covered with 30,922 silica-based tiles of varying composition, all firmly secured by adhesive.

OV-101 Enterprise made the first free gliding flight, after being carried aloft on top of a modified 747, on 12 August 1977. OV-102 Columbia made the first orbital flight, of 36 orbits, in April 1981. Subsequent first flights were: OV-099 Challenger, 4 April 1983; OV-103 Discovery, 30 August 1984; and OV-104 Atlantis, 3 October 1985. OV-105 Endeavor was ordered to replace Challenger, which was destroyed by SRB failure on 28 January 1986. The only other disaster came on 1 February 2003 when Columbia lost a refractory tile. This trivial damage resulted in the entire vehicle breaking up on re-entry.

In the Soviet Union, the leading OKB (experimental construction bureau) in the field of aerospace-planes was initially the famous one named for A. I. Mikoyan, popularly called MiG. On 29 June 1966 one of their senior designers, G. Ye. Lozino-Lozinskii, was named Chief Constructor of the Spiral project. There were several studies for this, but the definitive vehicle has yet to be built. The most promising configuration was a two-stage piggy-back system. The first stage, called 50–50, comprised a giant reusable carrier aircraft with four turboramjet engines in a box under the rear. The ogival wing had an inboard portion swept at 78°, integral with the fuselage and forming an arrowhead starting at the hemispherical nose, on the front of the single-seat cockpit. On the wingtips, at the rear, were endplate fins and rudders. Take-off weight with payload was to be 308,650lb. The payload, a single-seat aerospace-plane, was to separate at Mach 5.5, and then be boosted to orbital velocity by a tandem rocket for which S.P. Korolyev was responsible. Known as 50, the payload was to weigh about 22,700lb, and models designated BOA-4 and BOA-5 were flown on Cosmos missions. They were three-finned lifting bodies, with a flat underside and bulged top.

Lozino-Lozinskii's team did get funding to study an interim vehicle called EPOS (experimental piloted orbital aircraft). This again had the plan shape of a 78° arrowhead, but with a wing on each side with leading-edge sweep of 55°. Each wing could be pivoted up through 30°. Wing area was 71sq ft, and projected plan area of the

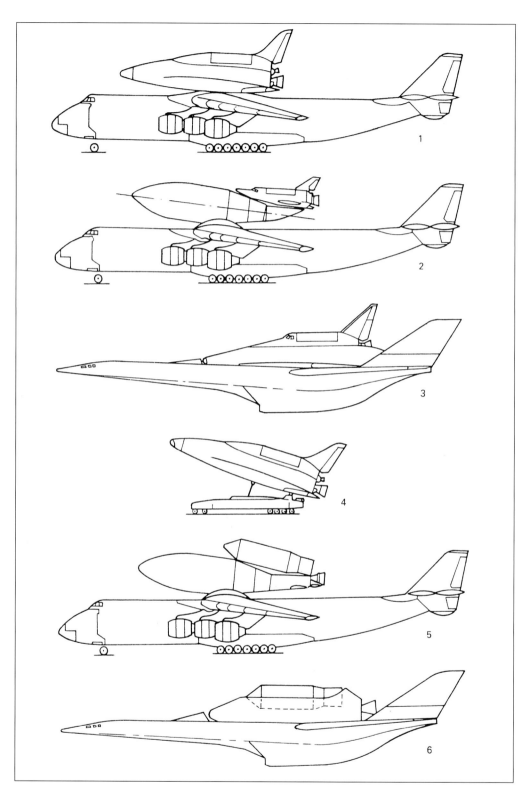

Artist's impressions of six possible arrangements for a Soviet aerospace-plane, explained in the text opposite. *via Philip Clark*

fuselage 258sq ft. At the back was a large highly swept fin and rudder, with airbrakes on each side. Elevons were on each wing. In the tail were the Saturn RD-36-35K turbojet, of 4,410lb thrust, fed by a retractable dorsal inlet, and six large and ten small control rockets, as well as a larger (3,310lb thrust) rocket for orbital manoeuvres and deorbit retro-thrust. The landing gear comprised four legs, each with a skid.

For preliminary testing, from 1974, the bureau built a non-orbital demonstrator, designated 105-11. This has essentially the EPOS airframe, and most of the systems, but differs in four major respects. The main rocket is absent, as are the attitude control thrusters. The FBW controls were replaced by traditional electro-hydraulic controls (Type ARS-40), the forward skids were replaced by wheels, and the cockpit contains conventional atmospheric-flight instruments. Test pilot A. G. Fastovets made the first free take-off on 11 October 1976, flying 19km from one runway to another (Ramenskoye). On 27 November 1977 he made a flight after being dropped from a Tu-95K at 16,400ft. The eighth and last flight took place in September 1978. Today this trim 9,300lb vehicle is at the VVS Museum at Monino.

This vehicle helped V. P. Glushko's team to develop the gigantic one-shot Energiya launcher for the Buran (snowstorm) aerospace-plane. After prolonged study, this vehicle assumed a configuration so like the US Shuttle orbiter that one has to look hard to tell which is which. Indeed, the wing area and dimensions of the payload were identical! Nevertheless, these vehicles differed from the US orbiter in important ways. They were launched mounted on the side of the awesome Energiya, with a core diameter of over 13ft. With four strap-on boosters, the total thrust at booster separation was no less than 8,900,000lb. Unlike the Shuttle orbiter, Buran vehicles had no main engines, only orbital manoeuvring engines and reaction-control thrusters. Another difference was that the boosters burned lox/kerosene. Because there were no main engines in the orbiter, its wing was further forward, but the twin-wheel nose gear was further back. The crew compartment was slightly larger, accommodation being provided for up to ten cosmonauts. The RCC leading edges and 38,000 ceramic tiles were very like those of the Shuttle orbiter, and the instrument landing system was also similar, but the Soviet orbiter deployed three braking parachutes after touchdown.

Span and length of the Buran were 78ft 9in and 119ft 5in, respectively, and lift-off weight was in the range 224,000 to 232,000lb, the whole assembly plus booster weighing about 5,370,000lb. In the initial programme, four Burans were being built. The first, partly dismantled to reduce weight to 40 tonnes, was carried to Baikonur Cosmodrome riding on the VM-T Atlant, a modified M-4 bomber. Subsequent Burans, complete, were conveyed by the enormous Antonov An-225 – which, thus loaded, caused a sensation at the 1989 Paris airshow, arriving like a fighter! The first launch, on 15 November 1988, was unmanned. After separation at 62 miles, two firings of the manoeuvring engines established a circular orbit at 155 miles. After two orbits re-entry at Mach 25 was made at an AOA of 40°, peak temperature reaching 1,535°C (2,800°F), followed by an automatic landing. Flight time was 206 minutes. With the collapse of the USSR, funding ceased, and on 12 May 2002 the manned flight orbiter was destroyed by the collapse of its hangar roof.

Computer graphics show six variations on the Buran theme, scaled up to the 250-tonne payload limit of the An-225. No. 1 shows the original version, but launched from the An-225. This would place 5 tonnes payload in a 124mile orbit at 51°. No. 2 shows an orbiter riding on a large external tank, launched together from the An-225. Payload would rise to 7 tonnes, or 8 tonnes if unmanned. No. 3 shows a hypersonic first-stage carrier in the 50–50 class, weighing 400 tonnes (650 with the upper stage). Engines would be the usual 88,185lb turboramjets, giving velocity at separation of Mach 5. Payload would be 25 tonnes. In No. 4 the 250-tonne Buran lifts off at 336mph from a rocket-powered trolley. A piloted version would have a payload of 1.5 tonnes. No. 5 shows the payload mounted on a one-shot expendable rocket, carried on the An-225, though this would increase cost per launch, the payload would rise to 20 tonnes. The last variant shows a similar one-shot second stage riding on the hypersonic carrier. Here payload would be no less than 37 tonnes, but again at the expense of high costs. Lozino-Lozinskii stressed the desirability of multinational cooperation in the development of the second stages, and told me confidentially that France was going to be a major partner.

We heard no more of Buran, until at the MAKS 2005 airshow in Moscow a full-scale mock-up was exhibited of a vehicle in this class to orbit a payload of 500kg. In February 2006 a derivative was announced, for operation

The 1990 configuration for France's Hermes. Attached to the rear is the MRH. *Aérospatiale*

by 2012, with Energiya, Molniya and Khrunichev participating. This is to 'combine features of Soyuz and Buran'.

Turning to France, in October 1985, the CNES (Centre National d'Etudes Spatiales) announced that Aérospatiale had been awarded the prime contract for a small Shuttle-type orbiter known as Hermès. Aérospatiale brought in Dassault to handle all Hermès work necessary for atmospheric flight. Subsequently the programme encountered many difficulties. Always an extremely modest vehicle, Hermès had by late 1990 been further downgraded to save weight. Originally planned to carry six persons, and have a Shuttle-type payload bay, it was redesigned to carry a crew of three and have no payload bay. To reduce weight further on reentry, many components and systems were moved out of the aerospacecraft and into a module de resources Hermès (MRH) attached at the back, which would be jettisoned just before re-entry, to burn up.

As eventually conceived, Hermès had a configuration resembling the HL-10, but without the central fin (an almost identical shape was BOA-4, tested by the Soviet Union on Cosmos flight 1445 on 15 March 1983). Span and length were to be 29ft 7in and 60ft, respectively, the length being reduced to 41ft 8in after jettisoning the MRH. Launch was to be by an Ariane 5 rocket, with two solid boosters added. The launch weight was estimated at 63,934lb, and re-entry weight 33,070lb. Maximum outbound payload was estimated at 3,527lb, which seemed unimpressive for a project optimistically costed at $2.6 billion.

France managed to rope in most of the EEC nations, the members of the customer European Space Agency (ESA), to help share the cost. The plan was that these countries would all agree to a go-ahead on the project in June 1991. This was expected to lead to the first unmanned subsonic flight in 1997, and the first manned orbital flight in 1999. Dassault-Breguet hoped to get a contract for a Falcon 900 trijet to be used for crew training. The launch vehicle was to be the Ariane 5, a giant rocket plus boosters which is totally expendable on each launch, leaving just the tiny aerospace-plane to be recovered. In view of this, Dassault-Breguet then proposed that Hermès should instead be launched by a gigantic aerospace-plane called Star-H (Système de Transport Aérobie Recuperable – Hermès). The Hermès would ride piggyback on the monster, which would have a similar configuration, with a delta wing curved up at

the tips into fins. Powered by five turbo-rockets, each rated at 88,185lb, Star-H would weigh 881,850lb at take-off. This was one of the occasions where the author felt justified in describing something as a pipe dream. He was right; in November 1993 the ESA announced that, as the USA and Russia could do any space-flighting necessary, Hermès would be terminated.

Perfidious Albion – as the French call us British when they find we have a mind of our own – was very publicly criticised by France for refusing to come in on Hermès. To do so would appear to have been ludicrous, because for many years Britain has had a far superior vehicle, Hotol, which needs no gigantic launcher and is fully recoverable. The only problem is that, true to form, the British government has done everything it can to ensure it never gets off the ground.

Hotol stands for Horizontal Take Off and Landing, which is what every sensible aerospace-plane surely ought to aim at. The concept was worked out from 1980 by British Aerospace at its Warton, Filton, Stevenage and Brough sites, in close collaboration with Rolls-Royce. The latter's RB.545 Swallow engine is the key to Hotol (see the author's *Rolls-Royce Aero Engines*, p.247). It takes off as a turbofan, burning liquid hydrogen. The intensely cold liquid is passed through a heat exchanger to cool the air rammed in at the variable-geometry supersonic ventral inlet. At Mach 5 at 85,000ft the engine changes to pure rocket operation, the inlet being closed and faired over, and the atmospheric oxygen replaced by liquid oxygen. In its original form Hotol was to have an aft wing of Concorde form, with a span of 68ft. Length was to be 196ft 10in, take-off weight 507,000lb, and payload into 162-mile orbit 17,635lb. Take-off was to be from a laser-guided trolley, lift-off occurring at 320mph In orbit, virtually every kind of space operation was to be possible. There was also to be an airline version, cruising at Mach 9, with the cargo bay replaced by a 70-passenger cabin. Its landing weight was to be 103,620lb, the landing run being 3,750ft.

By 1989 Hotol had been refined to have a pointed nose, with a smaller cone angle, a square-tipped delta wing of 93ft span, a length of 206ft 7in, and a take-off weight of 551,155lb. Unfortunately, not only had the government announced by this time that it would not provide funds, despite the willingness of the industrial partners to match government funding, but it also declared that the technology of the propulsion system was secret, thereby ensuring that no international partners could be found to enable the project to go forward. Despite this, BAe and Rolls-Royce were so convinced of the obvious rightness of Hotol, or something very like it, that from 1987 they continued to work on it with their own money. In early 1990 BAe announced that a way had been found to use a plain metal skin without insulating-tile protection, the upper surface (cooler on re-entry) being titanium. But how to go forward?

At the 1989 Paris airshow Ivan Yates, then BAe's Deputy Chief Executive, Engineering, took a cool look at the giant An-225, with the Buran on its back, and realised that, rather than merely carrying an aerospace-plane to the launch site, it could actually serve as the launcher. Talks with the Soviet Aviation Ministry and the Antonov bureau followed, and at the 1990 Farnborough show a big model was unveiled which caused gasps: a Hotol taking off from an An-225! The big plus was that, having been lifted to 30,000ft at 400mph, the Hotol no longer needed the variable-cycle feature burning oxygen from the air. This made it simpler, propulsion

Germany is making a real effort to get Sanger/Horus off the drawing board.

being just a high-pressure cryogenic rocket, of Soviet origin, and sidestepped the ridiculous 'secret' ban. The resulting vehicle, called Interim Hotol, would have four LO_2/LH_2 engines grouped in its blunt tail, would place about 15,500lb in a 186mile orbit, and cost an estimated $2.3 billion to develop. The big plus is that, apart from having five times the payload of Hermès, IH would cost about $8 million per mission, or one-third of the cost of using the Ariane 5 launcher (for Hermès or any other payload). If by any miracle the political obstacles could be overcome, and IH prove a commercial success, BAe hoped eventually to return to the original idea of using a dual-mode engine in a vehicle needing no outside help. Today, however, BAe has become BAE Systems, which has a hard-nosed attitude generally unfavourable to aerospace, except American aerospace.

Late in the previous century, Japan put a big effort into an aerospace-plane called Hope. Managed by Nasda (National Aero Space Development Agency), this was to fly supply missions to space stations, in partnership with NASA, and on independent sorties carrying laboratories and other loads in its Shuttle-style cargo bay. Initially to be unmanned, it was expected later to develop into various versions with a human crew, including a projected

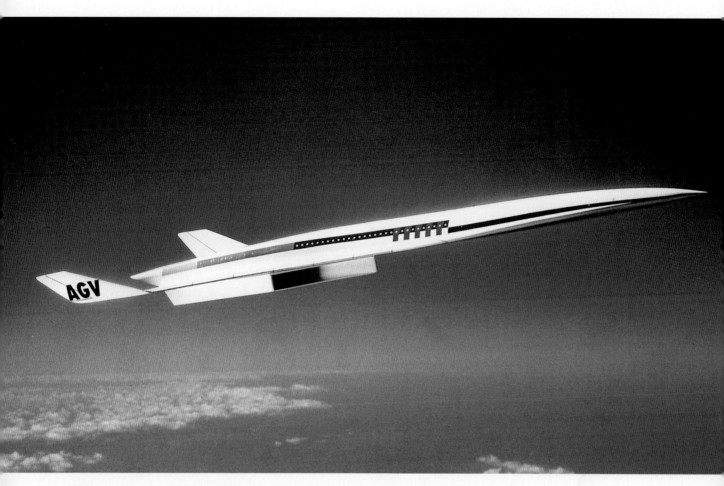

airline version. Missions were to last up to 100 hours, ending with an automatic ILS landing at 186mph the programme grew to involve several companies, notably Kawasaki. In shape, the vehicle had much in common with X-20 Dyna-Soar, but, as tends to happen, it grew in size and weight. By the time a very informative exhibit was displayed at the 1990 Farnborough airshow, even the original unmanned version had grown to 54ft 2in in length and 28,660lb in weight. This in turn meant that the H-II launch rocket would require not two but six added solid boosters. By 1991 H-II was more than a year late, because of explosions and fires during testing of the Mitsubishi LE-7 first-stage engine. H-II was then set for a maiden launch in 1993, with Hope payloads flying six years later. Hughes were sufficiently impressed to reserve 10 launch slots, but by 1998 the list of failures was impressive, and the US satellite company pulled out a year later. Budgetary pressure forced cancellation of H-II in 1999, and in 2003, instead of starting flight tests, the whole programme was abandoned (other

priorities had emerged, such as satellites to watch North Korean nuclear activity). Nasda then joined with two other agencies to form the Japan Aerospace Exploration Agency (JAXA), whose SST work is briefly noted in Chapter 9.

China is achieving increasing success in promoting its various versions of Long March rocket to launch customer payloads, and has for many years been planning to develop a national aerospace-plane, which could at any time be announced as a funded programme. Germany had for many years had a comprehensive Hypersonic Technology Programme which embraced a manned orbiter, Horus (hypersonic orbital re-usable upper stage), to be carried aloft riding on a giant ramjet-

Aérospatiale once made a big effort to get AGV off the drawing board. This was the 1989 configuration.

engined booster called, appropriately, Sänger. In 1990 this was halfway through a five-year system-definition programme funded at $190 million (a bit different from the British allocation of £375,000 to Hotol!). One of the main questions to be resolved was whether Sänger should take off with turbojets and then switch to independent ramjets, or use turboramjet engines in which air from a common variable supersonic inlet was diverted by flaps either through, or not through, the turbomachinery on its way to the ramjet, or whether the core engine could be allowed to windmill during ramjet operation. B burning LH$_2$, the ramjet went on test at MBB's Ottobrunn site on 7 June 1990, simulating up to Mach 4 at 20km (65,600ft). Sänger followed the usual shape except that its twin fins were mounted on the wide aft body. The small Horus, which had tip fins, was to separate from the Sänger at Mach 6.6, igniting its twin MBB high-pressure LO$_2$/LH$_2$ engines. This national programme was replaced by a 1994 decision to join the ESA on a project called Festip (future European space transportation investigative programmme), since when there has been an impressive amount of talking.

Like the Germans, the French are not afraid to look far ahead. At the 1989 Paris airshow Aérospatiale exhibited a beautiful model of the AGV (Avion a Grande Vitesse). They had spent years studying this 150-seater, which would cruise at Mach 5. By late 1990 they had rejected the original configuration in favour of

something with a shape more like Concorde. Meanwhile, Snecma, specialist in gas turbines, has joined forces with SEP, specialist in rocket engines, to create a propulsion system for what they call 'Hyperspace', and the AGV in particular. At the front is Snecma's turbofan machinery. Amidships are cryogenic turbopumps for LO$_2$ and LH$_2$, feeding the primary combustion chamber. At the back is the nozzle, arranged in sliding telescopic sections so that for Mach 5 at over 100,000ft it can be fully extended into a giant bellmouth (see artwork).

This leaves just the current American hypersonic programmes to be described in this book, and the story is, by no means for the first time, a lesson in how not to plan for the future.

It begins on a high note. In his State of the Union address of 4 February 1986 President Reagan announced the National Aero-Space Plane (NASP). This was to be a huge programme, linking some of the biggest names in the US aerospace industry. The President said that the Department of Defense had agreed to fund the NASP, via the Defense Advanced Research Projects Agency (Darpa) and the Strategic Defense Initiative, but he played down military applications. Instead

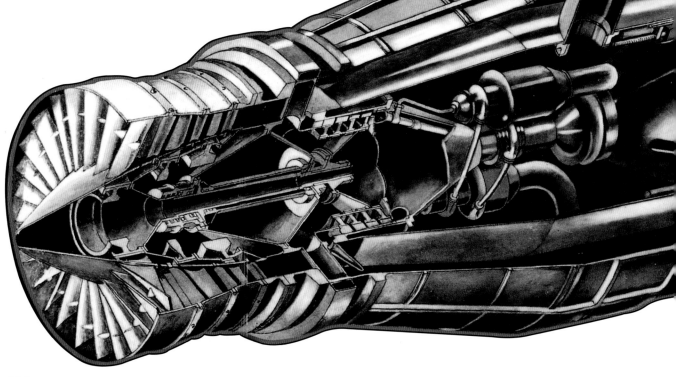

he concentrated on future hypersonic airliners, 'that could shrink travel times between Washington DC and Tokyo … to less than two hours'. Thus, the civil aerospace-plane became popularly called the Orient Express.

The initial programme, to create a test vehicle designated X-30, brought together ideas, materials, systems and design features collected by many agencies and companies since the Second World War. The basic concept was identical to Britain's Hotol, in that the pilot opens the throttles, the vehicle takes off down a runway, climbs out into space, accelerating to orbital speed, and eventually, at the end of its mission, returns to land back on the runway. This goal was challenging enough without worrying about precisely what civil or military missions might prove economically viable.

In April 1986 NASA and the DoD jointly awarded study contracts to seven companies, and in October 1987 awarded full study contracts to General Dynamics, McDonnell Douglas and Rockwell for the airframe and integration and Pratt & Whitney and Rocketdyne for propulsion. Every 'official' artist's impression of the X-30 published in 1990 showed a configuration submitted by Boeing, a loser! This configuration followed the norm in being an acute delta, with a pointed nose, but it did not have wingtips swept up to form fins. The wing was thin, and the fuselage enormous by comparison, and indicating the tankage needed. There was a single vertical tail, later replaced by a pair, while the central part of the wing extended aft in a flat platform. This was actually the upper edge of the propulsion system,

Hyperspace, the alliance between SNECMA and SEP, hoped to see this aerospace engine running on the test stand. A turbofan in the atmosphere, it becomes a cryogenic rocket in space. By 1990 the AGV had been almost completely redesigned, to a configuration closely similar to projected SSTs designed to fly less than half as fast! *Aerospatiale*

227

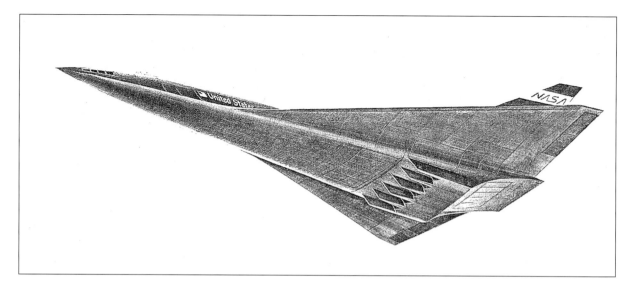

A 1977 Lockheed concept for a Mach 6 transport with a range of 5,750 miles. Five sets of inlet and nozzle doors show the dual-mode propulsion, with turbojets for take off and scramjets (supersonic-combustion ramjets) for cruise.

One of the configurations for the X-30 NASP, in this case by McDonnell Douglas in January 1989.

which was to comprise a row of three to five scramjets (supersonic-combustion ramjets) burning LH_2. These were to have variable inlets able to operate efficiently at Mach numbers from zero to 25. A main objective of the NASP was to make the forebody form an integral part of the inlet system, and the afterbody an integral part of the nozzle system. Rocketdyne was principal scramjet contractor, while Pratt & Whitney was being assisted by Marquardt, the pioneer of supersonic ramjets. For the airframe, in January 1990 the three principals formed a study consortium, Rockwell concentrating on titanium aluminide honeycomb sandwich, General Dynamics on carbon, and McDonnell Douglas on titanium reinforced with fibres of silicon.

The next phase involved building two X-30 flight vehicles. Each was to demonstrate aircraft mode, cruising at Mach 6 to 12 at 105,000ft, and spaceflight mode, accelerating to orbit at Mach 25. In the author's view optimistically, the first flight was scheduled for 1993. This was then revised to 'the late 1990s'. In fact, in May 1993 the whole programme was abandoned, because of 'budget cuts and uncertainties'. It has been replaced by numerous less challenging X-planes. These may eventually lead to a real aerospace-plane, in which a pilot will take off down a runway and fly clean out of the atmosphere at 16,500mph or more. He will probably look out through a glazed cockpit roof for all the world like Yeager's view from the XS-1, anything up to 100 years earlier. But, as this book suggests, we have – very haltingly – actually come a long way since then.

Rockwell 1990 artwork showing what was widely thought at the time to be the X-30 configuration that would actually be built.

INDEX

Aircraft

A-4 (V-2) 41-42, 45, 80, 93,
  197-198, 211
A.9/A.10 198-199
ACF (*Avion de Combat Futur*)
  163
Aerion Corporation 152
Aérospatiale 144, 147, 149
  AGV 148, 225-227
ASMP 112
ATSF 144-146, 148-149
  Hermès 222-224
AIDC Ching-Kuo 191
Airbus 35, 148
  A380 1
Ames Industrial Corporation
  AD-1 144
AMSA (Advanced Manned
  Strategic Aircraft) 117
Antonov An-225 215, 221, 224
Apollo lunar vehicle 204
Armstrong-Whitworth 111
  Scimitar 155
  Siskin ix
Arsenal de
  l'Aéronautique 69
AST (Advanced Supersonic
  Transport) 44, 147
ASTOVL 157
ATF (Advanced Tactical
  Fighter) 102-103, 188-189, 191
Avro 111
  Ashton 111
  Lancaster 29, 177
  Vulcan 111
  504 ix; 504N 19
  720 67, 111
  730 110-112, 116
  731 110-112
Avro Canada
  CF-100 64
  CF-105 Arrow 64-65;
    Mk 2 64

BAC (British Aircraft
  Corporation) 67, 125, 129-130,
  132-133
  Lightning F6 viii

TSR.2 111, 115, 132, 165
  221 136
BAE Systems 224
Bell x, 39, 44-45, 85, 93, 200,
  203, 208
  P-39 Airacobra (F2L) 93
  P-63A Kingcobra
    (L-39) 93
  X-1 (XS-1) x, 43-47, 80,
    83-85, 89, 91, 93, 204,
    211; X-1A 83-84; X-1B
    47, 83-84; X-1C 47;
    X-1D 47, 83-84;
    X-1E 85, 109
  X-2 (XS-2) 91, 93
  X-5 41, 119, 161
  58 (MX-984) 47
Blohm und Voss P.202 143
BOA-4 218, 223; BOA-5 218
Boeing 93, 113, 142-143,
  147-149, 155, 161-162, 167,
  176, 189, 228
  B-29 41, 45-46, 65, 83-84;
    RB-29 82; PB2B-1S
    86-87
  B-47 27, 107, 113; B-47B
    Stratojet 50
  B-50 83, 93; EB-50A 47, 93
  B-52 113, 167; B52H 22,
    191; NB-52 202, 205-206,
    213; NB-52B 212
  E-8 Joint-STARS 152
  E/A-18G Growler 177
  F-15 Eagle x, 172; F-15B
    152; F-15C 173; F-15E
    172; F-15S/MTD
    172-173, 187
  F/A18E 177; F/A18F 177
  F-111 99, 109, 119,
    161-162, 164, 166, 210;
    F-111B 161-162
  HSCT 148-149
  IM-99 Bomarc 80-82, 95
  KC-135Q 95
  SST 140-142, 168, 206
  XB-47D 50
  X-20 Dyna-Soar 207-208,
    210, 225

X-32B 159
707 214
727 143
733 140-142
747 197, 214, 218; 747-400
  148
2707 140-142
Boeing Canada 158
Bristol 125
  Brabazon 177
  188 xiv, 89-90, 111, 136
  198 SST 125
  223 125, 129
British Aerospace (BAe)
  147-149, 215, 223-224
  ACT Jaguar 182
  AST 148-150
  EAP 188
  Harrier 22, 153, 155, 157
  Hotol 215, 223-224, 226,
    228
  Interim Hotol 224
Buran 221, 223-224

CAC
  J-6 54
  J-10 188-189
Canadair Sabre 51
Chance Vought
  F-8 Crusader (F8U) 77,
    188, XF8U-1
  Crusader 77; XF8U-3
  Crusader III 77, 88
Concorde ix, xi, 19, 22, 27, 33,
  38, 97, 99, 103-105, 124-127,
  129, 132-136, 142, 144,
  146-148, 150, 168, 182, 223
Concorde II 144
Convair 58-59, 107, 214,
  218
  B-36 108
  B-58A Hustler 107-109,
    113, 117, 137, 162;
    B-58C 137; TB-58 109
  CV-880 137
  CV-990 137
  F2Y Sea Dart 79
  F-102 (Model 8-80) Delta

Dagger 30, 58-59, 61, 89,
  107; F-102A 61;
  F-102B 61
F-106 Delta Dart 61, 88,
  169, 191
SST 137-138
XF2Y-1 Sea Dart 79
XF-92A 59-60
YF-102 59; YF-102A (Model
  8-90) 62

Dassault 111-112, 130, 163, 223
  Balzac 154-155
  Mirage I 69; III 69, 71, 99,
    112, 154-155, 164, 183;
    IIIA 69; IV 112; IVA
    112; IVP 112
  Mirage IIIV 154-155
  Mirage G 163
  Mirage MD.550 70
  Mirage 2000 iv, 180, 183
  Mystère 69; II 69; IV 69
  Ouragan 69
  Super Mystère SMB.1 70;
    SMBB2 69
Dassault-Breguet 223
  Falcon 900
  Rafale 77, 98, 188-189
de Havilland
  DH106 Comet 43
  DH108 42-43
  Mosquito 39
  Sea Vixen 155
  Tiger Moth 48, 197
  Vampire 18, 150
  Venom 18
Deutsche Airbus 150
DFS 39, 41
  228 39
  346 35, 39-41
Dornberger/Ehricke BoMi 200,
  203
Douglas 58, 75, 85, 93, 200
  DC-3 16
  DC-8-40 75
  DC-10 134
  D-558-1 Skystreak 85-86,
    200

D-558-11 Skyrocket 48,
    80, 84-88, 93, 200-201
D-558-III (Dash-III,
    Model 671) 201-204
F-6A Skyray (F4D) 75, 77
X-3 90-91, 93
XF5D-1 Skylancer 75-76
Skyray 85
DVL 41

English Electric 65, 111
Canberra 19, 60, 64
Lightning 19, 64, 66-67,
    71, 99; F1 67
P.1 60, 66, 77; P.1B 66
EPOS (experimental piloted
    orbital aircraft) 218, 221
Eurofighter Typhoon 63, 101,
    188-189
EWR-Süd VJ 101C 155; VJ
    101D 155
EWR-Süd/Fairchild Republic
    AVS 155

Fairchild Republic 155, 214
Fairey 13, 65, 69
    F.D.2 65-66, 69, 136
Fieseler Fi 103 (V-1) 80
Focke-Wulf Ta 183 35
Fokker-Republic D.24 *Allianz*
FS-1 183

General Dynamics (GD) 137,
    189, 191, 228
AFTI F-16 182, 193-194
F-16 Fighting Falcon x, 33,
    98, 158, 176, 179,
    182-183, 191; F-16C 183
F-111A 111, 162
YF-16 173, 176, 182
YF-111A 162
Gloster
Javelin 29
Meteor 18, 27
Griffith Swallow 159
Grumman 214, 218
EA-6B Prowler 177
F9F Panther 18
F11F-1 Tiger (F9F-9) 76-77
F-14 99, 109, 160, 189,
    191; F-14A 6, 162
F-111B 162
XF10F Jaguar 161
X-29 192; X-29A 192-194
Guppy 93
Gulfstream Aerospace 151
Gulfstream II 151
QSJ 152
Gulfstream/Sukhoi SSBJ 150-151

Handley Page
HP100 110
HP.115 124, 136
Victor 75

Hawker 13, 18, 52, 153
Hunter 52, 155
Hurricane 10
P.1083 52
P.1099 52
P.1127 153, 155
P.1150 153, 155
P.1154 155
Sea Hawk 18
Tempest 13
Tornado 10
Typhoon 10-13, 15-16, 34
Hawker Siddeley 125, 128,
    161
Heinkel P.1068 39
Helios I 93
HSCT (High Speed Commercial
    Transport) 144
HSCT (US) (High Speed Civil
    Transport) 136, 148-149
Hughes 59, 226

Ilyushin Il-54 107
Israel Aircraft Industries Lavi
    188
JSF (Joint Strike Fighter) 159
Junkers Ju 287 191

Kawasaki 225

Lavochkin 52
La-160 53
La-174TK 53
La-176 52-53
La-190 53-54
Learjet 55LR 95
Leduc O.10 71; O.16 71; O.21
    71; O.22 71
Lockheed 15, 63, 139, 189, 191,
    228
CL-400 93-94
CL-823 140
CL-1200 Lancer 63
D-21A 95
F-80 18
F-104 Starfighter (Model 83)
    61, 72, 82, 88, 99, 109, 117,
    142, 155; F-104A 61-62;
    F-104C 61; F-104G 63;
    F-104S 63
F-117A 24, 183
L-1011-500 177
P-38 Lightning 15-16, 43;
    P-38F 15; P-38J-25 16
SR-71 Blackbird 1, 19, 22, 27,
    94-95, 99, 101-103, 137, 139,
    190; SR-71A 95, 117
U-2 89, 93
X-7 82-83; X-7A-1 82;
    X-7A-3 82
XF-104 62
XP-38 16
XP-80 15
XQ-5 82

Lockheed Boeing General
    Dynamics
F-22 Raptor 190, 194; F-22A
    191; FB-22 191
YF-22 x, 189-190
Lockheed-Martin
F-22 Raptor xii, 19, 64; F-22A
    34, 58, 98
F-35 Lightning II vi, 19,
    159, 161; F-35A 159;
    F-35B 155, 157, 159;
    F-35C 159
X-35B 159
YF-12A 94-95
LTV A-7 177

Martin
P6M-2 SeaMaster 75
X-24 (SV-5D) 213; X-24A
    (SV-5P) 211-213; X-24B
    (SV-5J) 212-213; X-24C
    214
MBB F-104CCV 182
McDonnell
F-4A Phantom II
    (XF4H-1) 77-79, 99, 101,
    109, 155, 172, 176-177,
    179; F-4CCV 182; F-4J
    78
F-101 Voodoo 58-59;
    F-101B 59
XF-88 Voodoo 49, 58;
    XF-88B 49
McDonnell Douglas 75, 150,
    176, 214, 228
F-15 1, 34, 79, 99, 109,
    186
F-18 x; F/A-18 Hornet 75,
    98, 173, 176-178,
    189, 196
McDonnell Douglas/
    Grumman A-12 74, 94, 137,
    139, 177
Messerschmitt 51
Bf 109 58
Me 163 18, 35; 163B 35,
    40, 42
Me 209 59
Me 262 18, 35; 262A-1a
    Schwalbe 35; A-2a
    Sturmvogel 35
P.1101 35, 41, 119, 161
Mikoyan/Guryevich (MiG) 52,
    72, 165
MiG-15 51, 61
MiG-17 53
MiG-19 54; MiG-19S 54
MiG-21 58, 72, 73, 99,
    164; MiG-21PF 73, 164
MiG-23 164-165;
    MiG-23PD 164
MiG-25 'Foxbat' x, 27,
    73-74, 95, 99, 172;
    MiG-25P 73-74;

MiG-25R 74
MiG-27 165
MiG-29 x, 99, 109,
    185-188, 191; MiG-29A
    186; MiG-29K 187;
    MiG-29UB 187;
    Baaz 187
MiG-31 x, 74
MiG-211 (MiG
    21MF/A-144) 135-137
SI (I-330) 53; 1-350 54
SM 53; SM-2 53-54;
    SM-12 54
Ye-152-1 (Ye-166) 99
Ye-155P 95
Ye-266 74
23-11/1 164
50 218-219
103-II 219
Miles 37, 39
Falcon 37
M.52 x, 36-39, 43, 48, 64
Ministry of Supply specifications
    and requirements
E.24/43 37, 39, 48
E.R.103 65-66, 89
F.1/43 18
F.18/37 10
F.124D 67
F.155T 65
R.156T 109
RB.156D 111
346 119
Mitsubishi T-2CCV 182
MRCA (Multi-Role Combat
    Aircraft) 162
Myasishchyev
M-4 221; VM-T Atlant 221
M-50 113
103-M (M-4) 113

Napier-Heston Racer 36
NASA
AAW F/A-18A 182
ACTIVE F-15B 182
M2-F1 209, 211-212
NASP X-30A x-xi, 80, 208,
    226-9
X-50 xi
Nasda 224
Hope 224-225
NBMR-3 153, 155
Nord 71
1500-01 Griffon I 71, 136
1500-02 Griffon II 71
North American (NAA) 57,
    113, 115, 193, 208
A3J-1/A-5 Vigilante 99,
    109-110; RA-5C 109
B-1 116-117, 166
F-30 51
F-86 Sabre 43, 47, 52, 57,
    88, 176; F-86A 51, 66; F-86D
    55; F-86F-25 51

F-100 (NA-180) Super
  Sabre 54-55, 73, 84;
  F-100A 54, 57-58;
  F-100B 57-58; F-100C 65
F-107 57-58, 88, 151;
  F-107A 77
F-108 Rapier 63-65, 74, 115
NA-134/FJ-1 Fury 51
NA-140 51
NA-157 55
NA-165 55
NAC-60 140
NR-356 157
P-51 Mustang 11-12,
  17-18, 35
SM-64A Navaho 80, 82,
  208
T-6 Harvard 13, 15, 48-49
X-10 80-81
X-15 (Project 1226) x,
  26, 33, 93, 113, 196-197,
  202-206, 208; X-15A-2
  204-207
XB-70 (B-70) Valkyrie 33,
  73-74, 95, 99, 107, 113,
  116-117, 119, 137-139,
  142, 144, 167; RS-70 117
XFV-12 156-158
XP-86 Sabre 51, 54
YF-100 57
North American Rockwell
  B-1A 167; B-1B Lancer 117,
  123, 167-171, 177
Northrop 75, 189, 191, 213
  Fang 75
  F-5 Freedom Fighter 75
  F-5E Tiger II 75
  F-18L 177
  F-20A (F-5G) Tigershark 75
  F-89 18
  HL-10 210, 212, 223
  MX-334 45
  M2-F2 212-213; M2-F3 213
  T-38 Talon 75
  YF-17 173, 176-177
Northrop Grumman B-2 183
Northrop McDonnell Douglas
  YF-23 x, 103, 189-190

OKB-2 41
  346P 41
  346-2 41; 346-3 41

Panavia Tornado 162-163; F.3
  162, 177

Republic 88, 140
  F-105 Thunderchief 63
  P-47 Thunderbolt 16, 43;
  P-47B 17; P-47C 16
  XF-84F Thunderflash 49;

XF-84H 49
XF-91 Thunderceptor 87-89
XF-103 xi, 88-89, 96
Rockwell 156, 193, 214, 228,
  230
Rockwell/MBB X-31A 193-194

Saab
  JAS 39 Gripen 181, 188-189
  J35 Draken 71-72
  37 Viggen 75; AJ37 75;
    JA37 75; SF37 75; SH37
    75; Sk37 75
  210 71-72
Sänger-Bredt antipodal
  bomber 199
Sänger/Horus 224, 226
Saunders-Roe (Saro)
  SR.A/1 79
  SR.53 67-68
  SR.177 67-68
SCAT (Supersonic
  Commercial Air Transport)
  138; SCAT-16 139-140;
  SCAT-17 139
SEPECAT Jaguar 75
SFECMAS 69, 71
  1402 Gerfaut 71, 136
  1501 71
Short SB.5 66
Shuttle orbiter 1, 80, 208,
  213-214, 216-218, 221
  OV-099 Challenger;
  OV-101 Enterprise 218;
  OV-102 Columbia 218;
  OV103 Discovery 218;
  OV-104 Atlantis 218;
  OV-105 Endeavour 218
Siebel Flugzeugwerke 39, 41
SNCASO
  SO.9000 Trident 68-69,
    136
  SO.9050 68
Spiral project 218
SSBJ (supersonic business jet)
  150
SST2 103-104, 149
Star-H 223
Sud Aviation 129-130, 132
  Caravelle 129
  Super Caravelle 129-130, 132
  Vautour 112
Sukhoi (Su) 72, 150, 164, 165,
  187
  S-51 150
  Su-7 73, 120; Su-7B 73
  Su-9 73, 164
  Su-11 73-74, 164
  Su-15 73-75, 99, 164;
    Su-15 T-5 74; Su-15VD 75
  Su-17 164

Su-24 99-101, 165
Su-27 99-101, 109, 127, 186,
  187-188, 190;
Su-27UB 187
Su-30M 187
Su-71G 164
T-10-21 188; T-10-24 187
Supermarine 18
  F.7/30 17
  S6 and S6B 13
  300 17
  Spitfire 15-18, 35; Mk VIII
    18; Mk IX 17-18; PR
    Mk XI 17-18; Mk XIV 18
  Spiteful 18

TFX (Tactical Fighter
  Experimental) 138, 155,
  161
Tsybin
  NM-1 89
  RSR 89
Tupolev 106
  Tu-20/142 'Bear' 48
  Tu-22M 120-121;
    Tu-22M-0 'Backfire' 121;
    Tu-22M-2 123;
    Tu-22-M3 99, 121, 123
  Tu-28 72; Tu-28P 72
  Tu-95K 219, 221
  Tu-98 72, 107, 117
  Tu-105 (Tu-22) 'Blinder'
    72, 107, 117, 120-121
  Tu-106 120
  Tu-128 72-73
  Tu-144 106, 133-137;
    Tu-144D 136
  Tu-160 'Blackjack' 99, 123,
    136, 166, 168-171

USAF
  requirement MX-1554 89
  SST 140

Vickers 39
  725 110
Vickers-Armstrongs 119, 125,
  161
  Swallow 119, 138, 161
  Wild Goose 119
Vickers Supermarine 111
  E.10/44 Attacker 18
  Swift 52
  Type 545 52, 146
Vought 214

Westland 47
Welkin 18-19
Yakovlev
  Yak-41 158; Yak-41M 159
  Yak-141 158-159, 164

Engines
Aerojet-General 208
  XCAL-200 45
Allison 155, 157
  J35 85
  XT38 49
  XT40 49
Armstrong Siddeley
  Adder 72
  P.159 110, 116
  P.176 111
  Sapphire 66, 77
  Screamer 67
  Viper 67, 69

Bristol 153
  Centaurus 13
  Olympus 89
Bristol Siddeley
  BS.100 155
  Orpheus 155

Chevrolet J35-C-3 51
Curtiss-Wright
  RJ47 80
  XLR11 89, 93
  XLR25 93

de Havilland
  Goblin 2 42-43
  Gyron Junior 68, 89
  Spectre 67; 5A 68
Dobrynin
  VD-7 113

Eurojet EJ200 189

Garrett/Honeywell
  F125 191
General Electric (GE) 108, 139,
  142
  F110 162, 172
  F120/YF120 102-103, 189
  F136 159
  F404 75, 177, 194
  F414 177
  GE4/J4C 140-141
  GE4/J5P 140
  J47 50-51; D-series 55
  J57 108
  J79 61-62, 77, 79, 108-109,
    115, 137
  J85 75
  J93 63, 115, 117
  MF295 162
  TG-180 (J35) 51, 85
  X279M 139-140
  YJ101 173

Isotov RD-33 186
Ivchenko D-18T 215

Koptchyenko R-79 187
Kuznetsov
  NK-22 121
  NK-25 123
  NK-144 120, 135
  NK-321 136, 171

Lyul'ka 106, 151
  AL-5 53
  AL-7 48, 107; AL-7F 107
  AL-31F 186
  AL-35F 187; AL-35N 188
  AM-5 53
Lyul'ka-Saturn 106

Marquardt 96, 228
  RJ43 82, 95
MBB 226
Mikulin AM-5 89
Mitsubishi LE-7 225

Napier 10
  Double Scorpion 67

Orenda Iroquois 64

P. F. Zubets 113
Power Jets W.2B/700 36-37
Pratt & Whiney 103, 228
  F100 172-173, 176
  F119/YF119 103, 159, 188, 191
  F135 159
  F401 157
  JT3 55
  JT8D-219 152
  JT11D-20 (J58) 94-95, 102
  JT12A 213
  J42 55
  J48 55
  J57 58, 61, 75, 77
  J58 (JT11B) 137, 139
  J75 58, 63-64, 94; J75B 112
  PW1120 188
  Wasp 13; Double 13; Major 13
  X-287 (JT9A-20) 139
  X-291 139
  304-2 93

Reaction Motors (RMI) x, 45
  6000C4 (XLR1 1-RM-3) 45
  LR8 85
  XLR8-RM-5 86
  XLR11 99, 206; XLR11-RM-5 (Dash-5) 47
  XLR30 201-202
  XLR99 204, 206
Reshetnikov D30F-6 74
Rocketdyne 80, 208, 218, 228

Rolls-Royce 10, 39, 65, 96, 103-106, 111, 147, 151, 153, 155, 159, 224
  Avon (AJ.65) 39, 52, 65-66, 71; Avon 200 52
  Derwent 5 52
  Griffon 18
  Merlin 13, 18
  Nene (RD-45) 52-53, 60
  Olympus 125, 132; Mk 320 111; Mk 593 103-104; Mk 621 132
  Tay 55
  Trent 215
  RB.106 52
  RB.108 155
  RB.121 110
  RB.145 155
  RB.153 119; RB.153-61 155
  RB.162 154-155, 161
  RB.545 Swallow 223
  XJ99 155
Rolls-Royce North America 157

Saturn RD-36-35K 221
SEP 227
SEPR.481 68-69
SNECMA 105-106, 133, 227
  Atar 112; 163; 9K 112; 101C 71; 101G 69
  MCV 99 105-106
  TF 106 155
  TF306 163
Soyuz R-79V-300 158

TsIAM RD-0120 215
Tumanskii 73
  R-27 164
  R-29B 164; R-29-300 165
  R-31 74
  RD-9B 53-54
  RD-36-35 165
  RD-41 159
Turbomeca
  Gabizo 68
  Marboré 68

Walter HWK 509B 40; 109-509C 41
Westinghouse
  J34 49, 79, 86, 90
  J40 80
  J46 79
Wright
  J65 50, 62, 77
  J67 89
  XRJ55 89
  YT49 50

People
Ackeret, 7, 35, 201
Adams, Mike 208
Allen, H. Julian 211
Apt, Capt Milburn G. 93
Armstrong, Neil 85, 208
Attinello, John D. 61

Beamont, R. P. 'Bee' 12, 66
Bell, Lawrence 44
Bennett, Don 142
Betz, Albert 7, 19, 35
Blériot, Louis 109
Blinov, Aleksandr 150
Bredt, Irene 199
Brennan Maurice 67
Bridgeman, Bill 86, 90
Brown, Cdr (later Capt) E. M. 'Winkle' 37, 43
Bulman, P. W. S. 'George' 18
Burcham, Milo 15
Büsemann, Adolf 7, 19, 35
Bylciw, Walt 103

Camm, Sir Sydney 10, 153, 155
Carter, President 167
Chemel, Capt Edouard 130
Chyepkin, V. 186
Clarkson, Lawrence W. 148
Cotton, Col Joe 117
Crossfield, Scott 206, 208
Cunningham, John 43

Dana, Bill 208
Dassault, Marcel 66
Davies, Capt 69
Davies, Handel 18
de Havilland Jr, Geoffrey 43
de Laval, G. C. P. 7
Derry, John 43
Donovan, Bob 85
Dornberger, Dr Walter 200

Eggers, Alfred J. 115-116, 211
Ehricke, Krafft 200
Engle, Joe 208
Erickson, Beryl 109
Everest, Col Frank 'Pete' 57-58, 84, 93

Fairey, Sir Richard 65
Farley, John 186
Fastovets, A. G. 221
Ford, President 167

Geddes, J. Philip 116
Glauert, 7
Glushko, V. P. 221
Goodlin, Chalmers 'Slick' 45
Goodmanson, Lloyd 143
Goryainov, 113

Gray, W. E. 125
Guignard, Jacques 68

Heinemann, Ed 75-76, 85, 88, 200-203
Hilton, Dr W. F. 1
Hooker, Sir Stanley 133, 153, 155
Hopkins, Capt Zeke 57
Hough, Lt-Col Cass 15-16
Hsueh-Sen, Dr C. 200
Ilyushin, Vladimir 74, 187
Ingells, Douglas J. 15-16

Johnson, Clarence L. 'Kelly' 15, 61, 63, 94-95, 102
Johnson, Dick 109
Johnston, A. M. 'Tex' 93
Jones, Robert T. 143

Kasmin, P. I. 41
Kelsey, Ben 15-16
King, Rex 66
Kittinger DFC, Capt Joe 34
Knight, Maj William 'Pete' 206, 208
Kokkinaki, Vladimir 107
Korolyev, S. P. 218
Kotcher, Ezra 43-44
Kracht, Felix 35, 39
Küchemann, D. 61, 126

Lavochkin, Semyon 53
Leduc, René 71
LeMay, Curtis 115
Lighthill, M. J. 18
Lipko, 113
Lippisch, Alexander 35, 39, 58, 75, 107, 161
Lockspeiser, Sir Ben 39
Lozino-Lozinskii, G. Ye. 218
Lyul'ka, Arkhip 48

Mach, Ernst 1, 7
MacKinnon, Malcolm 148
Mair, Flt Lt 18
Marquardt, Roy 96
Martin, John F. 86
Martindale, Sqn Ldr A. F. (Tony) 17-18, 35
Maskell, E. C. 126
Maxwell, James Clerk 197
McKay, John 208
McNamara, Robert S. 155, 161, 210
Messerschmitt, Willy 35
Meyer, 7, 24
Mikoyan, Artyom I. 53, 218-219
Miles, F. G. 39
Miles, George 37, 39
Missimer, William C. 103

Mitchell, Reginald 16-18
Morgan, M. B. 124
Multhopp, Dr Hans 35, 39
Murphy, Maj Elmer E. 109
Myasishchyev, V. M. 112

Nixon, President 142, 214
Noetinger, Jacques 52

Paulson, Allen 151
Peterson, Bruce 213
Peterson, Forrest 208
Petter, Teddy 19
Plenier, Jacques 147-148
Prandtl, 7, 24, 35

Quesada, Lt-Gen Elwood 'Pete' 138
Quill, Jeffrey 18

Reagan, President x, 167, 227
Rice, Ray 55
Rich, Ben 95
Richbourg, Charles E. 79

Robins, Sir Ralph 148
Ross, H. E. 199
Rushworth, Bob 208
Russell, Archibald (Doc) 130, 132
Rutan, Bert 144

Sandys, Lord Duncan 64, 66-67
Sänger, Eugen 199
Sarkisov, A. A. 186
Satre, Pierre 130
Sebold, Dick 107, 137-138
Servanty, Lucien 130
Shurcliff, William A. 142
Simonov, Mikhail P. 150-151
Slade, Gordon 65
Smith, George 58
Smith, Joe 18
Smith, R. A. 199
Sokolovskii, Capt O. V. 53
Stack, John 43-44, 61, 161
Stanley, Bob 85
Stanton, 7

Steiner, Jack 130
Stephenson, Air Cdre Geoffrey 58
Storms, Harrison 204
Strang, Bill 132
Swadling, Sid 148, 150
Swearingen, Ed 150
Syverston, Clarence A. 116

Taylor, G. I. 7
Thompson, Milt 208
Tobin, Sqn Ldr J. R. 17
Tsybin, P. V. 89
Tupolev, Andrei N. 72, 135
Twiss, Peter 65

Van Every, Kermit 85, 201
Vlasov, Col Evgeni 171
Voigt, Waldemar 35
von Braun, Wernher 198-199
von Kármán, Theodor 44

Walker, Joe 91, 206, 208
Waller, Ken 43
Wallis, Sir Barnes N. 39, 119, 138, 161
Welch, George 51, 57
Whitcombe, Richard T. 61
White, Bob 208
Whittle, Sir Frank ix-x, 19, 36, 199
Wilmot, John 39
Winter, Dr 39
Woods, Robert 44
Woolams, Jack 45, 93
Wren, E. A. 'Chris' 12, 15

Yates, Ivan 224
Yeager, Charles 'Chuck' x, 43, 45-46, 84, 231

Zagainov, Guerman I. 151
Ziegler, Jean 'Skip' 84
Ziese, Wolfgang 41

WITHDRAWN FROM STOCK WEXFORD PUBLIC LIBRARIES